After the Ark?

After the Ark?

Environmental Policy Making and the Zoo

Nicole A. Mazur

MELBOURNE UNIVERSITY PRESS

MELBOURNE UNIVERSITY PRESS
PO Box 278, Carlton South, Victoria 3053, Australia
info@mup.unimelb.edu.au
www.mup.com.au

First published 2001

Typeset by Syarikat Seng Teik Sdn. Bhd., Malaysia in 10 point Sabon
Printed in Australia by RossCo Print

National Library of Australia Cataloguing-in-Publication entry

Mazur, Nicole Andrea.
After the ark?: environmental policy making and the zoo.
Bibliography.
Includes index.
ISBN 0 522 84947 4.
1. Zoos. 2. Wildlife management. I. Title.
590.73

Contents

Illustrations

Plates

Figures

Tables

Acknowledgements

I WOULD NOT HAVE BEEN able to complete the research and subsequent work for this book without the assistance and support of so many people.

During the latter part of 1998 and for all of 1999, I undertook my work on zoos as a Reshaping Australian Institutions Postdoctoral Research Fellow at the Urban & Environment Program in the Australian National University's Research School of Social Sciences. I am indebted to Professor Patrick Troy, Dr Tim Bonyhady, Professor Max Neutze, Dr Nicholas Brown and Dr Brendan Gleeson for sharing their fellowship, critical knowledge and considerable academic experience with me. Penny Hanley tirelessly edited my work, Rita Coles was always on hand with technical and administrative support, and both of them provided invaluable moral support. My time at UEP was also enriched by the company of Christine Cannon, Dr Alice Roughley, Dr Peter Reid, Dr John Dargavel, Kurt Iveson and Coralie Cullen. Dr Annie Dugdale of the Sociology Department, Dr Tom Griffiths in the History Department, and Dr Anna Carr in the Centre for Resource & Environmental Studies read various parts of my manuscript and provided important and constructive advice. Dr Frank Castles of the Political Science Department was also highly supportive of my work and helped to ensure that I had sufficient time to complete the project.

I am indebted to Dr David Bennett, Executive Director of the Australian Academy for the Humanities, for showing me the 'zoo path' back in 1991, for never losing interest in my work and for always being willing to listen.

I would also like to thank the 'team' at Melbourne University Press—Brian Wilder, John Meckan, Andrew Watson, Margot Jones, Gillian Fulcher and Jean Dunn—for their enthusiasm and support for this book and for their help in seeing it through to its completion.

The research and analyses I conducted between 1993 and 1997 was undertaken at the Mawson Graduate Centre for Environmental Studies at the University of Adelaide. Dr Tim Doyle lent me his invaluable insights into environmental politics and provided me with his friendship and moral support. I am so much the richer for the time I shared with other close friends and colleagues at the Mawson Centre during that period: Margaret Cameron, Denise Noack, Elaine Stratford, Diane Favier, Patricia Carvalho, Stevie Austin, Natalie Harkin, Stephen Darley, Pam Keeler, Fiona De Rosa, Sue Dunn, Rob Ressom, Steve Baker, Adam Simpson, David Meiklejohn, Kris James, Dr Ken Dyer and so many others. Thanks also go to Drs Nick Harvey and Sandra Taylor who gave me important advice and direction during several stages of my doctoral research.

Dr Tim Clark of Yale University's School of Forestry and Environmental Studies and the Northern Rockies Conservation Cooperative has consistently taken time over the last several years to share with me his friendship, knowledge of the policy sciences and the importance that qualitative research has for addressing environmental problems. I would also like to thank Dr Stephen Kellert from the School of Forestry and Environmental Studies for his comments on a draft version of this book.

While I was conducting my research between 1993 and 1997, and again during 1998 and 1999, many members of the Australasian zoo community put aside time from their busy schedules to be interviewed by me (some more than once). Each conversation provided something unique to my research and I am especially grateful for those insights. Therefore, I would like to thank *all the people* I spoke to and to apologise to anyone who I might have left off this list. *Adelaide Zoo*: Rosalie Auricht, Susie Barlow, Bruce Campbell, Mark Craig, Graeme Craske, Phil Digney, John Draper, John Gardner, John Hopgood-Clark, David Langdon, Christina MacDonald, Ed McAllister, Jenny Moffit, Rob Morrison, Tim Roberts, Sheila Saville, Gert Skipper, Professor Mike Tyler, Peter Whitehead, the Friends of the Zoo. *Australasian Regional Association of Zoological Parks and Aquaria (ARAZPA)*: Christine Hopkins, Kevin Johnson, Kerry Kirk, Jonathan Wilcken. *Auckland Zoo*: Suzie Barlow, Wayne Boardman, Bruce (ungulate keeper), Chris Eastebrook, Ian (keeper), Richard Jacob-Hoff, David King, Jo Knight, Meagen McSweeney, G. Malcom, Martin (senior keeper), Maria, Chris (keepers), Laura Mumaw, Andrew Nelson, Olwyn (teacher), Trent (specialist keeper), Karen Turner, Peter West, Penni Whiting. *Currumbin Sanctuary*: Bruce Pascoe, Snezana Dudic, Dominic (teacher), Kerry Kitzelman, Helen Irwin, Michelle McAllen, Deryl McConahy, Rob McKinlay, Jeff McKee, Leon Misfeld, Nadya (research), Peter (veterinarian), Penny (wildlife presenter), Annette Pierce, Liz Romer, James Schwarz, Des Spitall, Jim Stockman, Trevor (keeper). *Healesville Sanctuary*: Rosie Booth, Michael D'Oleveira, Friends of the Zoo—Healesville, Justin Gamble, Greg Horrocks, Barry Krueger, Dianne Logg, Bruce MacDonald, Kevin Mason, Robert Mazzone, Pam Miskin, Julie Murphy, Scott Pullyblank, Luke Simpkins, Ian Smales, Kim White. *Melbourne Zoo*: John Arnott, Chris Banks, Annie Barker, John Birkett, Mike Brocklehurst, Peter Courtney, Cheryl Dean, Bev Drake, Amanda Embury, Pat Fielman, Friends of the Zoo—Melbourne Zoo, Merrill Haley, Judith Henke, Patrick Honan, Chris Hopkins, Greg Hunt, Chris Larcombe, Jane Leifman, Professor Angus Martin, Pam Parker, Richard Woods, Ron, Kerry, Sue (seal pool), Robert McAdam, Helen McCracken, Graham Mitchell, Peter Myroniuk, Lee Oosterweghel, Greg Pyers, Peter Rodgers, Greg Pyer, Gary Slater, Diane Smith, Peter Stroud, Stuart Tait, Peter Temple-Smith, Glenda Walker, Heidi Week, Paul Whitehorn, Jim Wilson. *Perth Zoo*: Dale Alp, Doug Anderson, John Arledge, Marcia Barclay, Peter Baker, Dean Burford, Allen Burtenshaw, Lynda Byart, Chris (ungulate keeper), Ian Crombie, John DeJose, Dick, Klaus, Rod, Shane (Endangered Species Unit), Carole Fuller, Reg Gates, Glenn Gaikhorst, Duncan Haliburton, Graham Hall, Neil Hamilton, Len Hitchen, Tony Hodge, Sherri Huntress, Colin Hyde, Tony Jupp, Jacqui Kennedy, Caroline Lawrence, Peter Leeflang, Charles McKinnon, Rosemary Markham, Daryl Miller, Julie Monument, Peter Mountford, Melanie Price, Laurie Pond, Helen Robertson, Glen Sullivan, Tom Tischler, Colin Wallbank,

Karen Wallbank, Tim Walton, Ian Williams, Lyn Williamson, Mike Yankovich, Zoo Volunteers. *Taronga Zoo*: Jane Bartos, Sylvia Bell, Mark Caddy, Polly Cevallos, Ken de la Motte, Kevin Evans, Karen Fifield, Gary Fry, Jack Giles, Wendy Gleen, Will Meickle, Larry (Australian mammals), Leanne Meirs, Libby (vet block), Steve McAuley, Frank McFadden, Darrill Miller, Dave Pepper-Edwards, Graham Phelps, Graham Phipps, Hunter Rankin, Angus Robinson, Russell (nocturnal house), Angela Schinman, Glenn Smith, Ron Strahan, Erna Walraven John West, De De Woodside, Zoo Friends. *Territory Wildlife Park*: Peter Cust, Gail (animal care), Gavin (birds of prey), Grant Husband, Karen (promotions), Leissa Kelley, Lee Moyes, Quentin (aquarium keeper), Robin (senior keeper), Kate Smith, Steve Templeton, Ida Williams. *Werribee Zoo*: Nick Atchison, Jenny Ford, Jenny Gardner, Tim Grech, Gaye Hamilton, David Hancocks, Peter Stroud, Larry Vogelnest. *Western Plains Zoo*: David Blyde, Peter Christie, Greg Horton, Ron (senior keeper), John Sainsbury.

I am grateful to Gary Backhouse, Rich Reading and Brian Miller for sharing with me their knowledge of policy and organisational processes in natural resource management and wildlife conservation. I had invaluable conversations with Margaret Rasa about the behavioural aspects of individual and organisational change.

Finally, I extend my heartfelt gratitude to my partner and confidante, Michael Martin, for his patience and to my family (Miriam Greene, Betty, Danielle, Alexandra, Peter and Helene Mazur) and friends (Claire Ciliotta, Nancy Loveland and Robin Sultzer) overseas for their long-distance support during my 'passage on the Ark'.

September 2000 *Nicki Mazur*

Abbreviations

AAZPA	American Association of Zoological Parks and Aquaria
ARAZPA	Australasian Regional Association of Zoological Parks and Aquaria
ASMP	Australasian Species Management Program
ASZK	Australasian Society of Zoo Keeping
AZA	American Zoo and Aquarium Association
CBSG	Captive Breeding Specialist Group
CITES	Convention of International Trade in Endangered Species
CRC	Cooperative Research Centre
IUCN	International Union for the Conservation of Nature
IUDZG	International Union of Directors of Zoological Gardens
NGO	non-government organisation
SCMZP	Sir Colin MacKenzie Zoological Park
SMP	Species Management Plan
SSP	Species Survival Plan
TAG	Taxon Advisory Group
TSCU	Threatened Species and Communities Unit
WPA	Wildlife Protection Authority
ZBV	Zoological Parks and Gardens Board of Victoria

Introduction

> *I shall cause a flood to rise over the land, and you must build an ark . . . In it you must make room for yourself and your family, and two—one male, one female—of every kind of creature, every kind of beast, reptile and bird. And you must fill it with enough food for them and for yourselves. For it will rain forty days and forty nights, and life on earth will be extinct.*
>
> Adapted from Genesis by Selina Hastings[1]

In the Bible, Noah eventually succeeded in rescuing a sample of all forms of life from the deluge. The flood waters receded, and Noah landed his ark and released the animals which then repopulated the scourged land. Over the last half a century, conservation biologists, authors and others have used Noah's Ark repeatedly as a metaphor. The Earth and various places on it have been depicted as arks which contain a diversity of life, but which must be managed in a particular way if their passengers are to be sustained through the dark waters of environmental degradation. In a similar fashion, zoo professionals have often labelled their institution as an 'ark', a vessel in which we could carry threatened species from around the world until the dangers to their habitats pass, at which time they could be released to replenish the Earth.

The mission of the modern Ark hinges upon the successful implementation of an internationally articulated and co-ordinated conservation policy. I define the zoo debate as a public policy issue —a social problem—and so seek insights into why debates about zoos' performance occur, and how these arguments can be effectively channelled into productive policy reform. Using this theoretical framework, I describe and explain how zoos' conservation policies are formulated, implemented and evaluated. Consequently, I set Australasian zoos in the international zoo policy context, drawing on the body of international zoo literature, and on North American and European zoo policy. I also use specific case material from research conducted at eight major Australian zoos and one New Zealand zoo. There are several reasons why the findings will be relevant to the international zoo community. Initially, European colonialism was instrumental in spreading a particular example of zoos across the globe. Later, North American zoo professionals had a strong role in developing the Western model of zoos. Most recently, the Australasian zoo community has brought its authority to bear in global zoo circles. And finally, there are certain scientific, administrative, economic and environmental trends that have made their mark on all Western institutions. The zoo is no exception.

Institutions and myths reflect the most deeply held values of societies. As such, the zoo is far more than a metaphor for a methodology of conserving wildlife or a collection of animals which demonstrates our fascination with the exotic. It is an institution that paradoxically reveals our anthropomorphic desire to see ourselves in animals while we simultaneously distance ourselves from them.[2] The zoo can certainly teach us about wildlife. Perhaps its symbolic value has even more to say about our relationships with non-human nature.

In recent times there has been an array of individual, organisational and institutional understandings of and reactions to ongoing and growing environmental problems. The zoo is one institution that has changed considerably in response to global environmental degradation. During the times of the ancient societies of the Greeks, Romans and Chinese, non-human nature in zoos was valued because it satisfied human curiosities and ambitions for power. By the middle of the nineteenth century, the zoo had become a public institution and touted science, education and recreation as its raison d'être. Over the last forty years, an international community of zoo professionals has recognised the need to 'help conserve Earth's fast-disappearing wildlife and biodiversity'.[3] Leaders in the zoo community, such as George Rabb, Director of the Brookfield Zoological Society, strive to move the zoo along an evolutionary continuum that will see it transformed from a 'living natural history cabinet' to an 'environmental resource centre'.[4]

These goals are ambitious and admirable. They are also disputed. People inside and outside the zoo community, internationally and in Australasia, are concerned about the modern zoo's aims and its capability for achieving them. There is a dearth of analysis on how decisions are made about zoos' priorities. Such 'policy analysis' is used to solve problems in many other areas, but has yet to be systematically applied to zoos. The large size and varied features of the zoo community means that zoos are expensive and difficult to study. In addition, the zoo community has often sought to defend itself in the face of its critics, some of whom would see the institution abolished altogether. Consequently, some zoos are protective of information about their practices. This has proven especially true for those zoos in Australia owned and operated by government agencies.

This book is predicated on the notion that the business of zoos is a complicated social process. Zoos are not merely collections of animals! They are as much about the people associated with them, as they are about the animals on display. The diversity of animal species in zoos is matched only by the numerous opinions about how zoos should fulfil their environmental obligations. Zoos' struggle to achieve their conservation mission reflects a much wider issue: namely, how we define and solve environmental problems.

Traditionally, environmental issues, such as the loss of biological diversity, have been defined as biological problems that require technical solutions. However, arresting or reversing any form of environmental degradation is just as much a public policy issue, one that requires us to spend more time 'managing' human actions than 'managing' nature—which often seems to defy our attempts at 'fixing' it.

No matter what institution they are found in, modern environmental concerns are not homogeneous. Environmentalism encompasses many different values, beliefs and perceptions. These views range from conservative, 'browns' to 'light greens', to radical, 'deep greens' and everything in between. This variety of positions means there will always be differences of opinion about how we should understand and respond to environmental problems. It is these conflicting definitions and the inherent difficulty and political risks associated with understanding and managing human behaviour and organisations that require the bulk of our attention.[5] However, our rigid institutional settings may make it difficult for us to achieve rapid and extensive change on this front. These matters suggest that the business of zoos is probably nothing short of complicated.

A substantial proportion of the current rhetoric surrounding the role of the contemporary zoo is predicated on how much the institution has changed. I begin, therefore, by setting the zoo's transformation in its historical context. How was the zoo transformed from a private menagerie to a public institution focused on conserving biodiversity? Distinctive social, political and economic factors would have shaped the zoo's evolution and spread a particular model of zoo practice across the globe. In formulating appropriate and feasible goals for their institution, today's zoo policy-makers must work in conditions that are similar to those operating in other cultural, scientific and educational institutions. They face a host of practical and ideological challenges resulting from contemporary political and administrative trends.

Having established the evolutionary context of modern zoo policy, I then consider the four main pillars that are commonly used to categorise zoo activities: conservation, research, recreation and education. The Ark's passenger lists and manifests are decided by species management programs, which make up the *ex-situ* (not in the animal's habitat) conservation function of zoos. These programs are used extensively by zoos to sustain their own collections and to contribute to conserving biodiversity. Yet these methods are also widely criticised on the basis of their particular design, how applicable they are to complex ecological problems and how well they are implemented within organisational settings, as well as at local, national, regional and international levels.

In response to logistical difficulties and criticisms about captive breeding's limited capacity to restore endangered species, zoo professionals have attempted to extend the zoo's responsibilities to include more *in-situ* (in the animal's natural habitat) conservation projects. Zoo professionals are increasingly looking to research to help them achieve this aim. I look at some of the social, institutional and political forces that shape scientific knowledge and the ways it has been used in traditional and contemporary zoo research programs. In some cases, these factors have helped the zoo community to make some important advances, in others they are holding the zoo back from making more significant conservation gains.

In addition to research, contemporary zoo professionals have stated that one of their most important missions is to facilitate public understanding of and concern for wildlife and broader environmental matters. In order to maintain regular and increasing revenue, zoos must also provide enjoyment for their visitors. Do these imperatives conflict with one another? The answer may lie partly in how 'education' is defined by zoo professionals, and how they use it to achieve environmental conservation. Zoo educators may face special ideological and administrative situations as they try to fulfil their modern goals. Examining the educational experiences which zoos offer provides insights into how well the Ark is helping to enlighten society.

In discussing the nature and efficacy of zoos' conservation, research, education and recreation programs, I also consider zoos' organisational arrangements and management ideologies. Western society's predilection for bureaucratic arrangements and cultures tends to foster conventional, conservative management ideologies and decision-making and to empower a managerial élite. What implications are these trends are likely to have for zoo policy? I describe some contemporary management practices on the Ark and the bureaucratic settings that characterise the larger public zoos in Australia. If the zoo's capacity to achieve its public good goals is restricted by its administrative arrangements, then hope for improvement will lie in examples where certain zoos have identified these dysfunctions and generated progressive regional and local conservation planning mechanisms.

The way the Ark is managed and financed is also probably critical to its success. In an age of economic rationalism, Western governments are increasingly unenthusiastic about fully supporting government-owned enterprises, and are pressuring museums, universities, hospitals and libraries to become more efficient and financially independent. Under this particular management framework and its fiscal arrangements, these institutions are being encouraged to adopt an increasing range of corporate efficiencies in achieving 'financial viability', and have turned to the private sector for funding. The manifestation of these trends in the zoo may set up special tensions, where zoo staff try to reconcile conflicts between the ecological imperatives of conservation and education programs and the economic mandates of commercialisation.

If zoos are to be likened to Noah's Ark, they may be sailing into uncharted waters. Zoos' success in conservation may lie in their capacity to further incorporate more progressive ideologies such as ecology into their principles and programs. Yet achieving such substantive environmental policy reform requires that professionals (and the citizenry alike) recognise and then challenge both long-standing and current assumptions of their institutions.

When I was undertaking the original research for this book, I was frequently asked 'Which zoo is the best', or 'What is the best zoo you have been to'? I often found myself replying, 'I do not know!' I did not set out to answer those questions in the first place, nor am I now attempting to identify 'good' or 'bad' zoos. I do not provide extensive technical information that hands down a last verdict or a universal law on zoos, nor do I spend time breaking zoo programs down into smaller and smaller variables, which can be quantified and then used to rank zoos in order of their performance. I do use some quantitative analysis, such as looking at the proportion of endangered species represented in zoo collections relative to more common and exotic animals, but I do so in order to understand *broad* trends in zoo policy.

In essence, I hope to contribute to improving the zoo's conservation role by discussing the *quality* of zoo policy. The book therefore questions the commonly-accepted and officially-defined goals of zoos and looks for the deeper meanings that are found in the human actions and events that make up zoo conservation rhetoric and programs. The way we communicate with one another generates certain fundamental principles that we then use to create rules about how society should conduct itself. In addition to highlighting the importance of these social relationships for zoos, the book seeks for contradictions that create tensions and conflict among individuals or among different groups of people and for information on how those conflicts are resolved (or not).

I am hopeful that this book will widen the scope and impact of the zoo debate by including some issues that have—to date—been neglected, even ignored altogether. There have been virtually no comprehensive analyses of the conservation role of major zoos in Australia, and very few in other countries; nor have the parallels between the practical and ideological challenges facing zoos and those facing cultural and educational institutions been highlighted. Perhaps most significant is the lack of inquiry about the development and implementation of zoos' conservation policies within the broader context of environmental values and policy-making processes.

This book is fundamentally oriented towards 'problems'. That is, I offer constructive criticism by highlighting particular policies from a selection of zoos that either fall short of or are inconsistent with their stated conservation aims and which therefore restrict the zoo's evolution. I also highlight examples of policies which could be or are being used to realise the conservation goals that the zoo community itself has so strongly called for. People who feel that the zoo's evolution is on the right course, and that its achievements should only be commended, may feel that this kind of discourse is confrontational and promotes unnecessary and unfounded faultfinding. Others, who think that there is always more that the zoo (and other institutions) can do for social and environmental problem-solving, will not necessarily agree with everything they read in this book. However, in the hope of bringing about positive change, they are likely to welcome the chance to discuss any problems—old or new—which might be inhibiting the zoo's evolution.

1

Assembling the Ark

He saved out Noah and his family, and arranged to exterminate the rest. He planned an Ark, and Noah built it. Neither of them had ever built an Ark before, nor knew anything about Arks; and so something out of the common was to be expected. It happened.

Mark Twain[1]

The conservation role of the zoo is much younger than the institution itself. Private menageries have a long history, beginning in ancient societies and have existed in just about every corner of the world. Far from being benevolent, collecting animals has been akin to displays of pomp and power, demonstrating the influence of the ruling classes whose members could obtain exotic animals from far-off lands. While it might be argued that the basic tenet of zoos is still to maintain animals in captivity, there have been some discernible changes in the philosophy and physical characteristics of many zoos. These transformations have always been closely linked to certain political and economic forces and shifts in human relationships with non-human nature during major historical periods. By the nineteenth century, a new form of public zoo was born in Europe. The seeds of today's zoo—conservation, science, education and recreation—were being planted and distributed to other parts of the world.

I shall focus primarily on the development of zoos in Europe, North America and Australasia. This is not meant to negate the importance that other regions have had for historic and contemporary zoo policy. Indeed, the modern Ark has a truly global presence.

Throughout Latin America, Africa and Asia, there are hundreds of zoos, which are visited by millions of people per year. There are also regional and national associations, which formulate and implement policies for these zoos. Nevertheless, Europe and North America have played the most central role in shaping the modern conservation role of the zoo. During the last several decades, Australasia has also become a key participant in global zoo policy-making. Finally, since there are some standard features to the form and policies of zoos which transcend geographical boundaries, valuable insights can be still gained from analysing zoo developments in select regions.

Ancient Menageries

Efforts to tame, domesticate and keep animals in captivity are nearly as old as human society itself. While the exact point in history when zoos began has been debated, most concur that collecting animals and maintaining menageries dates to the ancient societies of the Egyptians, Greeks, Romans and Chinese. The main purpose in maintaining these animal collections was to display the wealth and power of the owners and provide for exotic hunting and entertainment. Animals would also have been exchanged as gestures of diplomacy or friendship. For some members of these societies, particularly for the Egyptians, and in Mesopotamia, there would have been significant religious reasons for keeping animals.

Unlike their modern counterparts, ancient menageries did not have systematic collections plans. While some animal selection was motivated by human curiosity about new creatures discovered during exploration, war and conquest, an important rationale was the symbolism of power. Choosing animals for collections was often informed by the major symbolic value afforded to particular species as well as by the power conferred to the owner who possessed those specimens. Birds of prey, carnivorous mammals and charismatic mega-vertebrates such as elephants were well represented in most ancient menageries. While the Chinese and Greeks did confer some scientific and educational value onto their menageries, such acknowledgement was largely in its infancy and did not reappear until much later.

The justifications, if any, for maintaining vast collections of animals were no doubt informed by certain attitudes towards non-human nature. In Greek society, an Aristotelian perspective asserted that humans' capacity to reason placed us at the top of a hierarchy of nature whereby those beings with less reasoning ability exist for the sake of those with more (for example, plants for animals and humans, animals for humans). The advent of Christianity would magnify this hierarchical relationship and increase the gulf between humans and non-human nature by conferring onto people dominion over nature, and relegating animals to a place outside humans' sphere of concern.[2]

Roman civilisation epitomised these values. Extensive tracts of land were cleared for pastoral economies, and wildlife populations were devastated by gladiatorial games. Considerable energy was devoted to extending and defending the Roman Empire. The slaughter of animals and people in the gladiatorial games provide evidence that Romans were averse to bestowing moral consideration to weak beings. Instead, these creatures were transformed, en masse, into commodities of power. Even when the Roman Empire collapsed and such practices waned, the narrow parameters of care did not shift. On the contrary, people were more concerned that their declining civilisation testified to their backward transformation into a state of 'animality'.[3]

The Middle Ages

The records concerning human attitudes towards and treatment of non-human nature in the Middle Ages are rather sparse. It is possible to ascertain the predominant influence of Christian anthropocentrism in Europe. Vast changes to the environment such as clearing forests, draining land and creating gardens and hunting parks, were informed by the belief that humans should assist God in improving their earthly home.[4] Christian attitudes towards non-human nature were highly ambivalent.[5] Some individuals promoted treating non-humans with some care (such as St Francis of Assisi), but most treated animals with a mixture of pity and contempt. Civilisation itself reflected human's superiority over all other beings, which were considered to be 'beastly' or 'uncivilised'. Certainly, the traditions of maintaining menageries and hunting indicate that utilitarian and superior attitudes towards nature were fundamental to these times.

Exotic animals in medieval and later Europe were seen as luxurious gifts and prized possessions to be exchanged among royalty and important nobility. Many monarchs and aristocrats maintained private animal collections (or menageries) which featured 'ferocious' animals such as lions, leopards, tigers and bears, although the selection and abundance of animals varied according to particular preferences of the reigning monarch. Sport and entertainment were often equated with the hunting and/or fighting of 'wild beasts'. Some rulers fought their captive animals in an effort to demonstrate their triumph over the natural world.[6]

Some may have been motivated by less destructive interests. Frederick II, the Holy Roman Emperor and King of Sicily, has been depicted as an outstanding naturalist.[7] His vivarium, purported to be the first of its kind in Western Europe, contained a large area of marshes and ponds populated with species of water birds, and he produced a book on ornithology. Frederick II's interests, however learned they may have been, were probably not an indication that people appreciated the inherent value of non-human nature. Until the late Middle Ages, people studied nature because it was evidence of God's existence and His plans for a designed world, rather than of its inherent worth.[8]

Discovery and Trade during the Renaissance

By the Renaissance, menageries were a widespread custom not limited to European society. The Ming Dynasty of China gathered great collections of animals. During the great Ming Expeditions of 1405–1433, Zheng He brought back the first giraffe seen in China.[9] French travellers visiting the Arab and Turkish societies during the sixteenth century found expansive menageries and a long tradition of suppying menagerie animals to other parts of the world. In 1519, Cortes viewed the vast collection of Montezuma, ruler of the Aztecs.[10]

The colonisation of the New World, and discoveries of new environments, climates, flora and fauna continued to reinforce notions of nature's God-given fullness, richness and variety, and stimulated a great interest in learning.[11] Such erudition, however, was set in a context of exploitation. In this period habitat destruction was widespread throughout the Americas and Europe. These exploitative endeavors were probably facilitated by the humanist world view of the Renaissance. This ideology increased the gap between humans and non-humans by contrasting human superiority, dignity and potential with the more limited capacity of 'lower' beings. In conjunction with the anthropocentrism being extolled by contemporary Christianity, such a supremacist perspective contributed to an overall deterioration in the treatment of animals, even though some prominent figures promoted compassion for non-humans.

Hunting practices, such as the establishment of deer parks, provide some insight into human-animal relations in Europe at that time. Animals were confined in large forests and other areas set aside for the express purpose of hunting expeditions, a central feature of the royal and noble lifestyle. A scarcity of game in England after the fifteenth century led to a proliferation of these parks where large numbers of animals met their end. Many aristocrats throughout Europe possessed deer parks as well. Crown Prince Maximilian is credited with establishing the first modern menagerie in 1552, because he stocked his Deer Park with exotic animals for display purposes only.[12]

European discovery of the New World and a growth in trade between certain European centres and other parts of the world facilitated this period of zoo development which was also characterised by an increase in menageries.[13] Rare species from every part of the known world were being imported at an unprecedented rate to private menageries. Animal collections continued to symbolise colonial conquest, wealth and status and aesthetic satisfaction. Overseas discoveries and successful trade brought about increased levels of prosperity and invigorated longings for luxury goods. With the rise of the middle class, a new and numerous class of people now possessed greater resources, status and leisure time to cultivate those activities (including an interest in animals and nature) that had previously been confined to élites.

Renaissance Italy exemplifies these trends. Venetians played a key role in actualising the interest in animal keeping.[14] Venice served as the primary channel for movement and storage of goods. Hence, it was necessarily an important focus for the supply of wild animals to Italian menageries. By the sixteenth century all the great European cities, other Italian courts, several popes and cardinals, and private citizens kept menageries.

Animal collections were well established in European society at the end of the sixteenth and during the seventeenth centuries. In addition to the existence of royal menageries, most aristocrats maintained collections of resplendent birds in their gardens. The public was developing an interest in viewing exotic animals, and commercial exploitation of these interests grew. Societal values regarding non-human nature were changing as well. The anthropocentric views of the late Renaissance began to emerge, where concern was shown for the treatment of animals and preserving at least some wild creatures.[15] The seventeenth century also marks the beginning of an association between the scientific study of natural history and zoos.

Nonetheless, these trends had yet to become fully manifest in the practice of maintaining animal collections. Perhaps the influence of thinkers like Descartes still coloured contemporary thought. Descartes' notion was that human rationality and purpose should dominate non-human nature which was devoid of consciousness. In any case, most architectural designs of menageries emphasised pleasing visitors rather than meeting the needs of captive animals, as few knew about or were interested in animal behaviour.[16] For example, despite Louis XIV's menagerie having a reputation for serving scholarly interests at the time it was established more attention was devoted to its aesthetic architectural features than to the adverse conditions for animals in their small brick and iron enclosures. It was not until the mid-eighteenth century that definitive association of science and zoos began, and it was much later before animal welfare concerns started to influence the design of zoos.

Revolutions in Thought and Other Ark Developments

Significant streams of thought and particular attitudes towards nature emerging during the eighteenth and nineteenth centuries had marked effects on the principles and practices of zoos. The ideology of scientific reason was becoming a central focus for educated Europeans. Knowledge of natural history grew quickly and took on a particularly utilitarian character. Disciplines such as botany and zoology began as a means of identifying how plants and animals could benefit humankind. A growing concern for animal welfare and the works of Linnaeus, Darwin and others did challenge human distance from and treatment of non-human nature, but these ideas took a long time to take hold in zoos.

Increasing levels of affluence, geographic mobility, leisure time and mechanisation brought on by the Industrial Revolution, encouraged the appeal of 'wild' nature and a strengthening of the animal welfare movement. The middle classes would now have the time and money to pursue such interests. Interestingly enough, while these forces gave rise to an interest in conservation, they also contributed to an increase in the popularity of zoos. By the mid-nineteenth century, a distinctly European model of *public* zoos was spreading quickly across the globe. Here too are the earliest uses of science and education as justifications for zoos' existence, and the beginnings of the identity problems that confounded—and continue to confound—zoos' attempts to achieve the cultural status and legitimacy enjoyed by museums.

The French Revolution was a critical catalyst for the transformation of zoos from private to public institutions and for the incorporation of science into zoos' raison d'être. This was a time when 'ordinary' people called for the demise of many court frivolities and excesses, not the least of which were the menageries at Versailles. There was considerable indignation in certain circles over the fact that animals in menageries were living a life far above the standards of average citizens. Although specific details of the exact events vary, it is commonly agreed that disgruntled citizens, demanding the demise of such a vulgar indulgence, stormed the Versailles menagerie. Some of the animals were liberated, while others were slaughtered. The remaining dangerous animals and a few others were eventually transferred to a site adjacent to the Muséum d'Histoire naturelle. It signalled a new phase in zoo development. As the Museum and the Zoo were located at the same site, scientists had a counterpart 'museum' of living animals for study, and the people of France had a zoological garden from which they could not be excluded.

The Paris Zoo succeeded in achieving a high-profile scientific identity for itself. In effect it became the centre of biological research for France for a period of time. Many leading scientists were affiliated with the Paris Zoo and Natural History Museum and believed that the Zoo was more than a place where the 'ignorant' could (or should) stare at exotic beasts in a somewhat stupefied state. While citizens could not be prevented from viewing and enjoying the animals, the collection was always to be maintained, first and foremost, in the name of pursuing scientific knowledge. The Zoo and the Museum, like the menageries that preceded them, were under the control of powerful men of affairs who could shape policy. They were adamant that the living part of the Museum would not be mistaken for the flourishing menageries and trained animals shows, which they believed to be ordinary and frivolous. These early signs of conflict between the idea of the zoo as a place of entertainment and recreation versus the zoo as a place of substantial scientific and educational achievement would resurface many times over the next two centuries.

The success of the Paris Zoo was a significant catalyst for the establishment of the Zoological Society of London and London Zoological Gardens, as well as other zoos in Europe. The Jardin Des Plantes' indisputable reputation attracted the attention of somewhat envious British officials, particularly Sir Stamford Raffles, a distinguished colonial administrator in the East Indies. An avid animal collector—he possessed a substantial menagerie of his own —Raffles was inspired by the Paris example and wanted to see a similar institution established in London. He found the quality of menageries that existed in and around London to be lacking, believing they were designed to titillate the curiosity of the multitudes rather than to celebrate the achievements of a few. Such a serious national scientific and political deficit needed attending to. Raffles' activities as a naturalist seemed to mirror his concerns as a colonial administrator:

> He made discoveries, imposed order, and carried off whatever seemed particularly valuable or interesting. The maintenance and study of captive wild animals, simultaneous emblems of human mastery over the natural world and of English dominion over remote territories, offered an especially vivid rhetorical means of re-enacting and extending the work of empire, and Raffles intended to continue his colonial pursuits in this figurative form.[17]

Raffles was instrumental in actualising his vision of a London-based collection of living animals for scientific purposes and general interest. He collaborated with Sir Humphrey Davy and Sir Joseph Banks, influential men in their own right, to establish the Zoological Society of London in 1826. The Society envisaged offering 'a collection of living animals such as never yet existed in ancient or modern times' where animals would be gathered 'from every part of the globe to be applied either to some useful purpose, or as objects of scientific research, not of vulgar admiration'.[18] These sentiments would reappear many times over with the subsequent proliferation of other zoos throughout the remainder of the nineteenth and the first half of the twentieth century.

The example of the London Zoo also provides some valuable insights into what was to become a widespread and enduring scientific and educational conservatism in zoos. By the mid-nineteenth century, Linnaeus' system of classification for plants and animals was having a profound effect on both the science of zoology and the organisation and interpretation of zoological collections.[19] Zoos became a focal point for the emerging science of zoology by virtue of the animals they held or could hold (both alive and dead); the race was on to see who could identify and name the most specimens, or collect the most unique animals. The zoos of the late nineteenth and early twentieth century were characterised by their 'postage stamp' collections and their focus on taxonomic interests in zoology and biology. Lord Zuckerman describes how, despite its access to the most current scientific knowledge of the day, the

London Zoo resisted exploring new areas of zoological science. The Zoological Society of London remained exclusively committed to a straightforward study of anatomical form and classification. It has been suggested that most zoos lost their scientific integrity when the science of zoology moved away from anatomy.[20] Supposedly scientists exhausted the supply of knowledge that could be gained from studying zoo animals. The life sciences were now the new focus of concern, but had not been developed sufficiently to contribute to managing live animals in captivity. Hence, there was little incentive for 'serious' scientists to remain involved with zoos.

The Pursuit of Legitimacy

By the middle and late nineteenth century the zoo of today was becoming well established. There is little doubt that zoos were a popular destination, particularly for the middle classes. It became a public right to have access to a zoo and, at the very least, zoos were to be opened to a group of subscribers or to people paying an entrance fee. This growth has been linked closely with a sense of civic pride. Groups of prominent citizens perceived the need for *their* city to have a zoological garden—as well as a museum, library and art gallery—which appeared to be the way cities were afforded a sense of permanence, wealth and metropolitan identity.[21] Zoological gardens could now be found in cities all around the world: Buenos Aires, Helsinki, Moscow, Bombay, Tokyo, Dublin, Bristol, Amsterdam, Frankfurt, Basle and several cities in the United States, Africa and Australia. European expansion was instrumental in spreading a particular model of the zoological garden.

Acclimatisation activities at this time were critical to the establishment of Australasian zoos and promoted a kind of biological imperialism. Early settlers initially saw the flora, fauna and landscape of Australia as impoverished and dangerous, something that would impede their advancement.[22] Consequently, they set out to supplant the natural order by destroying indigenous nature and importing, collecting and releasing many wild and domestic exotic species for agricultural, recreational and aesthetic purposes. In Australia and New Zealand, Acclimatisation Committees or Societies co-ordinated most of these activities. Such organisations also promoted a hunting culture that depicted Australian fauna as too tame and therefore inferior to the big game trophies of Africa.[23] Members intent on improving the hunting stock endorsed the importation of species such as antelope, African eland, zebra and ostrich.

The Committees formed the foundation of municipal zoological gardens in Australia, in part by establishing processes for locating, transporting and caring for the animals they imported.[24] Like the European zoological societies, these organisations were constituted entirely by leading citizens and holders of public office who acted as managers and benefactors. These influential members lent an air of respectability to the Societies, and their power enabled them to 'fast track' intercolonial and international transactions in animals and money. E. J. L. Hallstrom, President of Taronga Trust for twenty-five years, spent an estimated one million pounds on Australian zoos and on Taronga Zoo in particular.[25] He funded various overseas expeditions, animal purchases and building projects. Numerous other personalities and benefactors were instrumental in establishing and maintaining a particular pattern of zoo development in Australia—consistent with that in other countries—over the course of their history.

Generally speaking, the middle and late nineteenth century was a significant period of zoo development. Zoos were proliferating and becoming a favoured leisure destination among the general public in most Western societies. Accompanying this trend were the beginnings of the international zoo community's contemporary justification for the existence of zoos: conservation, education, research and recreation. The founders of early zoos were declaring that their organisations were being established for the purpose of science and education. Zoos' transition from strictly élite to more publicly-oriented institutions may have created the need for offering such defences. In effect, zoos were moving into a period whereby they would need to be accountable to the public. These organisational goals were forming in the context of particular ethical, economic and political conditions.

The animal welfare issues that continue to plague zoos today were growing in strength during this time. Consideration for animals had intensified considerably. Humane advocates were active across the Empire by the 1840s, although most of the animal welfare activity was centrally located in England.[26] The Royal Society for the Prevention of Cruelty to Animals was founded in 1824 and the Royal Society for the Protection of Birds was set up in 1889. These developments would have created serious repercussions for the growing international zoo community. Anti-cruelty campaigns resulted in the enactment of significant pieces of legislation: the consolidation of the *Prevention of Cruelty Act (1849)* and the *Wild Animals in Captivity Protection Act (1900)*.[27] The latter reform was implemented in response to animal welfarists' dissatisfaction with the austere conditions for animals in zoos and circuses.

It was becoming hard to ignore detrimental effects that inadequate menagerie-like conditions were having on animals such as the big cats. Behaviours such as neurotic pacing or incessant weaving were displeasing to watch. There was also a growing interest in and desire for more naturalistic enclosures. Life expectancies for many zoo animals were poor: the average life span for the lions, tigers, leopards and pumas at the London Zoo was a dismal two years.[28] This also had serious financial and public relations ramifications for the Zoo, as the carnivores were the most popular attractions for those times.

Exhibit designs did not necessarily reflect the physiological or psychological needs of the animals they contained. Rather, as in the earlier royal menageries, wild and dangerous animals were confined in small cages that allowed for optimal viewing from the vantage point of clearly marked and manicured park pathways. Nineteenth and early twentieth century zoos were constituted by 'conspicuously attractive edifices designed in whatever was then the current fashionable style' and were essentially 'places for outings in pleasant surroundings'.[29] Additionally, the use of taxonomic classifications to organise the animal collections and exhibits was a way of imposing order onto non-human nature, which was seen as unruly or chaotic.

Notable changes to zoo design began when Carl Hagenbeck opened his zoo in Stellingen near Hamburg in 1897 and introduced husbandry practices of the times. Hagenbeck, a well-known animal trainer and trader, would have an enormous influence on international zoo practice. While his designs may not have escaped symbolising humans' dominion over non-human nature, they offered a way of presenting captive animals that was unique for those times. Hagenbeck wanted to recreate some semblance of the animal's 'normal' life and reduce perceived barriers between the observer and the observed. Visitors to his zoo saw animals in large, open-air naturalistic enclosures. Grassy ditches or water-filled moats replaced many of the bars. There were extensive trees and shrubs, as well as winding pathways to each exhibit. Some exhibits were arranged so that it appeared as if carnivores and herbivores, predator and prey were in virtually the same exhibit.

Hagenbeck also used his experience as an animal trainer to improve contemporary animal management practices in zoos. He emphasised differences between individual animals, the need for positive reinforcement and that training captive animals had occupational and psychological usefulness (for the animals). Until this time, animals from warmer climates were housed in airtight, heated buildings in the fear that exposing them to temperate climates would kill them. Unfortunately, these methods did little to preserve the animals' health. Hagenbeck was the first to dispense with such methods, showing the zoo world that, within reason, animals could adjust to climatic changes.

In comparison to Hagenbeck's interest in dispensing with cruel treatment of and inappropriate housing for wild animals, the commercial aspects of his profession would seem to be in direct contradiction to extending care to non-human nature. The growing number of zoos had created a market for animal specimens, as it was now no longer feasible to build a collection on gifts from fellow rulers and loyal subjects. Exotic animals had become 'a kind of currency'.[30] Hagenbeck was noted for his architectural and husbandry innovations, as well as for the size of his animal trading business. His business supplied hundreds of thousands of animals to zoos in Australia, America, Africa and Asia and circuses in America and Europe.

It is sobering to contemplate the effect that this kind of trade would have had on the wild populations of these animals given that

Hagenbeck was only *one* of *many* animal traders operating businesses around that time. One meagre defence for Hagenbeck is that his animal trade would not have had as dramatic an effect on the environment as farming and sporting practices of the day, or the subsequent World Wars.[31] It remains that zoo practices of that time would have made serious impacts on populations of wild animals, irrespective of any contrasts that might be made with other land-use practices. Zoos would eventually have to account for their practices when international regulations in wildlife trade were introduced during the second half of the twentieth century.

Nonetheless, there were early signs that zoo policy was shifting in response to conservation concerns, which were growing during the late nineteenth century. In Australia, the typical colonial perspective that had been hostile or indifferent towards the natural landscape began to shift towards an appreciation of the uniqueness of Australian natural and cultural heritage and the pragmatic use of natural resources.[32] By the mid- to late 1800s, conservation-minded citizens were increasingly concerned about the harmful effects that introducing exotic plants and animals were having on the fragile Australian landscape. Some of the most troublesome species in Australia today were already wreaking havoc. For example, rabbits numbered in the thousands and were displacing the smaller marsupials, pigs were thriving in and degrading the bushland, and weeds (horehound, thistles, and sweet briar) were dominating land that settlers had cleared.[33] Not surprisingly, some biologists in the scientific community were becoming increasingly interested in the preservation and study of indigenous flora and fauna.

This growing conservation consciousness, as well as some economic concerns, had a definitive impact on early Australian zoos. As awareness of the problems associated with introduced species grew, the Acclimatisation Societies eventually switched their emphasis from releasing exotic animals to focusing more on maintaining and developing their zoological gardens. Economic matters may also have triggered this policy shift. Catherine de Courcy has described how the Acclimatisation Societies, often organised by volunteers, were increasingly burdened by the expense of purchasing, maintaining and distributing animals.[34] In Melbourne, Albert

Le Souef began collecting Australian fauna when the Zoo was in serious financial trouble. Local animals were easy to acquire, cheap to maintain, and could be used as exchange currency with national and overseas zoos and acclimatisation groups.

The early conservation movement also appeared to be influencing zoo policy in America. The National Zoological Park is purported to be the first zoo established in America dedicated to the preservation of native species, as well as to the 'advancement of science and the instruction and recreation of the people'.[35] The New York Zoological Society also touted the conservation banner when it was chartered as a non-profit organisation. It endeavoured to provide an educational and recreational experience for the public through exhibits, research programs and conservation efforts, and identified wildlife conservation as one of its most foremost objectives.

Yet there was also some cynicism about the benevolent endeavours of these early American zoos, comparable in many ways to the anti-zoo sentiment emanating from the conservation community today. This doubt was embodied in a debate between the United States Senate and House over whether funds should be allocated to a national, scientifically oriented zoo.[36] The Senate was largely supportive of such an endeavour. The Congress, however, appeared to reflect popular sentiment by questioning the need for a zoo which would have to import exotic species and 'lock up' native species.

A New Era

During the early part of the twentieth century conservation interests were gradually building. Western environmental values at this time were focused largely on 'wise use' concepts in resource management.[37] These views were characterised by an instrumental world view of non-human nature, which attributes worth to conservation on the basis of benefits it provides to humans. Wildlife conservation efforts emphasised the prevention of species extinctions, but had yet to consider the welfare of individual animals.[38]

Several associations for wildlife protection were established. By 1900 Britain, France, Spain, Portugal, Germany, Italy and the Belgian Congo signed a Convention for the Preservation of Animals, Birds and Fish in Africa to curb the decimation of 'game' species. In 1903 the Society for Preservation of Wild Fauna of the Empire (Fauna Preservation Society) was founded and focused on species facing extinction.

Despite these developments, there were few signs during the early 1900s that conservation had infiltrated zoos to any greater degree than during the latter part of the previous century, nor had scientific or educational programs advanced very far. These trends are not entirely surprising, given that zoos were operating in a context of political and economic upheaval surrounding the two World Wars. Extensive personnel, material and monetary shortages in zoos would have severely hampered their efforts to maintain and upgrade exhibits. Many zoos fell into a state of disrepair. Zoo professionals began to rely on crowd-pleasing activities such as chimpanzee tea parties, elephant rides and other animal shows to attract visitor revenues. This reliance on entertainment programs to service commercial objectives would resurface in the mid-1980s and 1990s, when zoo professionals faced another period of economic difficulties.

Despite the economic malaise and a preference for commercialised practices, other developments were having a significant effect on zoo policy. Two influential associations were established, signalling a growing formalisation of the international zoo community: the International Union of Directors of Zoological Gardens (IUDZG) in 1935 and the American Association of Zoological Parks and Aquaria (AAZPA) in 1924. New ideas about zoos were realised through the opening in 1931 of the Whipsnade Zoo in England, seventy miles outside London. The Zoo was inspired in part by Hagenbeck's achievements. Chalmers Mitchell, the Secretary of the London Zoological Society from 1903–1935, believed that there was a need for a country property where animals would be presented in groups living in 'natural' surroundings. These conditions would foster normal social behaviours, particularly the breeding patterns necessary for maintaining a constant supply of exhibit animals, that could not be reproduced in the confined city zoo spaces. Additionally, animals from the London Zoo could be sent to Whipsnade for periods of rest and recuperation after illness. In the late twentieth century this model of operating dual city-country properties has become a critical component in the management of several Australian zoos, such as the Western Plains Zoo in Dubbo, NSW and the Monarto Zoological Park outside of Adelaide, South Australia.

The Modern Ark

The latter half of this century has been a formative period of consolidating and popularising both modern environmentalism and the role of zoos in conservation. Thinkers such as Rachel Carson have ignited Western society's concern about the harm that humans are causing to the biosphere. The cumulative effects of industrialisation and increased material consumption by an expanding population were considered to be serious problems. Awareness of mass species extinctions was building. Simultaneously, a post-war regeneration had launched zoos—now major recreational attractions—into another boom phase. Subsequent zoo redevelopment during the 1960s and 1970s incorporated new philosophies and program priorities as well as exhibit designs.

One area of scientific study to have a significant impact on zoo practices was ethology. The study of wild animal behaviour grew in popularity during the second half of the twentieth century. Heini Hediger's publication of *Wild Animals in Captivity* in 1950 was an early indicator of a growing awareness that zoos needed to be more responsive to animals' physiological and behavioural needs in captivity. Hediger had expanded upon Hagenbeck's animal training and behavioural work and laid the foundations for applying the science of Ethology in zoos.

By the late 1960s and early 1970s, zoo veterinarians and comparative psychologists were using Ethology to find ways to foster animals' natural behaviours in captivity and eradicate the neurotic, stereotypical acts that visitors saw so often. It was hoped that such advances would benefit not only the animals, but would also foster a better understanding of animal behaviour amongst the viewing public. Zoo-related biological sciences such as animal ecology and veterinary medicine for exotic animals, conspicuously absent during the first half of the century, made marked advances during the 1970s and 1980s. An especially important finding was the identification of stress as a threat equal in impact to disease. Many exhibit designs now incorporated flight distances: the amount of space an animal needs to retreat from approaching creatures in order to feel safe.[39] Additionally, field studies highlighted the intricacy of wild animals' social relations, and zoos began to apply these findings to the captive situation in order to enhance animal breeding and the overall welfare of species in their care.

Zoo exhibit designs continued to evolve during this time. By the end of the early 1970s most Western zoos had 'naturalistic' moated exhibits similar to Hagenbeck's original concept. However, the 'nature' animals were set in was often being simulated with high-tech materials. Corresponding interpretive materials comprised large graphic images and interactive gadgets. Naturalistic habitats were also a part of the landscape immersion concept introduced into zoos in the early 1980s. Here, designers strove for even more realism by placing the visitor, in effect, into the animal's environment.

The open-range zoo format was increasing in popularity as well. In these zoos, animals are kept in open paddocks simulating their 'natural' surrounds. The Western Plains Zoo in Dubbo, New South Wales, was eventually opened in 1977 after Henri Hediger's review of Taronga Zoo's operations in 1965 which recommended, among a series of structural changes, that the Taronga Trust acquire land outside Sydney to establish a facility similar to Whipsnade Park. The appeal of these open-range zoos lies in the fact that people can view animals close-up in quite spacious naturalistic surroundings, simulating the experience of encountering animals in the 'wild' without the associated dangers.

Despite these significant design shifts, most of the older zoos today exhibit a patchwork of the total range of design trends. This type of planning has been described as a kind of disjointed incrementalism:

> An irregular pulse of funding leaves its mark on the [zoo] landscape: exhibits that represent design eras separated by long gaps in time. Consequently, zoos in transition typically contain a disjointed set of exhibits. Older exhibits and facilities stand as conspicuous artifacts of the past, while newer, more naturalistic exhibits begin to surround them. It's not surprising that redeveloping zoos are often full of architectural and thematic collisions, as well as transitional path systems and detours around exhibit construction projects.[40]

By the latter half of the twentieth century education in zoos was becoming more complex. Theories of visitor behaviour and interpretation were applied to the zoo context so that zoos' potential to provide informal learning experiences for visitors could be developed further. Yet most zoo programs focused on formal schools education programs, directed at primary pupils in particular. There were several indications that traditional approaches to education would be redefined in the coming years. By 1966 the *International Zoo Yearbook* featured a section titled 'Zoo Education'. In the late 1960s the International Association of Zoo Educators was formed by several zoo education officers in Europe who were interested in examining the various problems and potential of zoo education. By the late 1970s the Association's membership included American and Australian representatives, and dialogues centred on whether zoo education should be focused merely on schools' curricula or whether it should also promote an appreciation of wildlife.

In the 1980s zoo education departments began to weave more environmental themes throughout program formats in an effort to wake 'the sleeping giant' of environmental education.[41] By 1982 ecological interpretation and nature conservation were the predominant themes at the International Association of Zoo Educators' Conference, and 'Communicating for Conservation' was the focus in 1988.[42] In addition to an earlier emphasis on communicating messages in order to create an appreciation of species and their habitats, zoo professionals now sought to motivate people to act on behalf of conservation from some understanding of the critical role our species plays in ecosystems locally and worldwide. Schoolchildren were (and remain) an important focal point for formal education, although zoos began extending their brief to include the wider community.

The Move to Conservation Principles and Practices

One of the more telling ways in which zoos responded to growing environmental and animal welfare concerns in the late twentieth century was to strengthen their resolve to become conservation-oriented organisations. Throughout North America, north-west Europe, Australia and New Zealand, zoo professionals were questioning the purpose of their institution and asking what they should achieve in the future. Many zoo staff members realised that the consumptive practice of continually replenishing their collections with specimens from the wild was no longer feasible. Stocking policies would now have to consider the shrinking supply of wild animals, and organisational and exhibit philosophies would need to reflect a changing environmental ethic. Displaying exotic animals for its own sake no longer provided sufficient rationale for zoos' existence, and husbandry practices had to account for individual animals' well-being.

These conditions created the perceived need to formalise the zoo community and elevate organisational performances. This is a time when the notion of 'good' and 'bad' zoos appeared. 'Good' zoos, it was contended, would be those that provide quality conservation and education programs, 'bad' zoos would not make such provisions, instead keeping their animals in inadequate enclosures. Several regions formed professional associations as a response to growing criticism. For example, the Federation of Zoos in Great Britain and Ireland was established in 1966. The aim of the Federation was to alleviate the concern of animal welfare groups by raising zoos' operating standards, and to this end, to lobby for legislation which would impose compulsory inspections for its member institutions. The legislation was delayed by an opposing rival organisation representing safari parks, but was finally actualised with the passing of the *Zoo Licensing Act* in 1981. The establishment of the Association of Zoo Directors of Australia and New Zealand in 1967 signalled the Australasian zoo community's perception of the need for a fraternity among the region's zoos. For the first thirteen years of its existence, the Association of Zoo Directors functioned primarily as a directors' association. By the time it had evolved into the Australasian Regional Association of Zoological Parks and Aquaria (ARAZPA), it was being used to promote and facilitate more committed, co-ordinated and scientific approaches to zoo management, acquisition, disposition, exchange, breeding and conservation of the region's zoo animals.[43]

Across the globe, there were calls for zoos to rethink their strategies. The Zoological Society of London launched the *International Zoo Yearbook* in 1959, a publication that would provide information on progress zoos were making in animal husbandry and management and display techniques. The Yearbook signalled the zoo community's push to achieve scientific legitimacy and has acted as a barometer for detecting the nature of international zoo activities. The first volume carried this emphasis:

> The need to encourage conservation of remaining wild fauna becomes more urgent every day . . . it is in the interests of everyone that the need to conserve wildlife and the move to develop zoos should be reconciled . . . this will be possible only if we ensure that zoos are designed and maintained in such a way that the wild animals housed in them not only live long in captivity, but also live in conditions where they can reproduce themselves.[44]

The times dictated that zoos maintain self-sustaining animal populations and increase the social relevance of their collections. The International Zoo Yearbook's first Conservation Section featured three articles highlighting developments and opportunities for zoos and conservation in 1962. By 1977 this section was now entitled 'Breeding Endangered Species in Captivity' and contained twenty-two articles on problems particular to breeding endangered species, genetic management issues, applying behavioural studies to captive breeding strategies, and demographic models for managing captive populations.

The realisation that nature would not offer an unending supply of wild animals for zoo collections was manifest in the growing emphasis on conserving endangered species as an integral component of zoos' programs. Gerald Durrell's Jersey Zoo was one of the earliest efforts to highlight the relevance of captive populations of endangered species for wildlife conservation. His animal collection was oriented specifically, and exclusively, for breeding endangered species for eventual reintroduction to the wild. He created the Jersey Wildlife Preservation Trust in 1963 in order to ensure the survival of this park. This organisation went on to become a powerful force in the international zoo community. Unencumbered by the commercial imperatives of many municipal zoos, it has become renowned for its capacity to maximise the utility of captive breeding schemes for reintroduction programs.

The first international symposium on the role of zoos in conservation was held in 1965. Zoo professionals saw a need for an international organisation that would create a policy on distributing rare animals for exhibition. The International Union for the Conservation of Nature (IUCN), IUDZG and the International Council for Bird Preservation sponsored this event. The event was attended by thirty-nine delegates from the three sponsoring bodies as well as those from AAZPA, the Fauna Preservation Society, the Zoological Society of London, the Institute of Biology, World Wildlife Fund and observers from several animal welfare non-government organisations, government game departments and commercial animal dealers. The symposium featured sessions on breeding endangered species in captivity; import, export, transport and sale of wild animals; conservation education in zoos; and moral and financial support for conservation through zoos. Additionally, the symposium examined the implementation of governmental controls on the importation and transit of rare animals.

These conventions were organised in the context of a particular political climate that led to the signing of the Convention on International Trade in Endangered Species of Wild Fauna and Flora (CITES) in 1972. The implications of international and national wildlife trade regulations were—and still are—sensitive issues for zoos. Zoo professionals wanted to be seen as an *active* partner in this push for stricter wildlife trade regulations, but they were also quite concerned by them. The spirit of the zoo community's concerns are captured in an excerpt from an article appearing in the *International Zoo Yearbook* in 1980:

> How does this global convention affect zoos; will it put an end to the purchase of rare animals and, more importantly, could it curtail the growing and vital exchange of captive-bred animals between zoos . . . One thing is certain: CITES not only multiplies the red-tape and paper work, which could discourage many zoos from attempting to overcome this challenge, but also it will, and must, affect the whole concept of breeding endangered species—the why, the how, and the what nows.[45]

These issues were no less significant for the Australasian region. There was a discernible absence of exotic fauna in Australasian zoos and of Australian fauna in overseas zoos due to the region's strict regulations, which were designed to prevent the introduction of diseases and feral pests. During this time no importation of birds was allowed and numerous restrictions existed for many other kinds of animals. AZDANZ consulted with and advised Australian Commonwealth authorities on the suitability of overseas public zoos to hold Australian fauna, and on other matters. These dialogues presumably would have included zoos negotiating for policies which were in their best interests; namely trying to ensure that regulations had a maximum impact on the illegal wildlife trade and a minimal effect on zoos' ability to exchange animals, or occasionally collect them from the wild.

By the 1970s Australasia's co-operative species management programs were forming in response to several factors. First, native species were growing increasingly rare or were simply difficult to obtain from the wild. Second, quarantine restrictions on importing many species constrained access to stocks that might otherwise be imported directly from certain disease-free countries. Third, the complete ban on importing birds highlighted the need to carefully manage existing stocks to ensure future displays. Finally, unwieldy logistics and prohibitive costs associated with importing larger mammals ensured that zoos found it too difficult to feature these animals. Several studbooks were created for high profile zoo species (giraffe, zebra, hippo and cassowary) and for species known to be declining in Australasian collections. There was some breeding and collection planning for priority species and an annual census begun for other species. Generally speaking, however, extensive schemes for breeding stocks did not begin until the mid-1980s.

In the earlier part of the 1980s there was a growing realisation that the viability of zoos' contribution to conserving endangered species increasingly hinged upon the pooling of efforts, as no single institution had either the genetic or organisational resources necessary to guarantee the long-term survival of a species. By the mid-1980s the emphasis on breeding endangered species in captivity was at its peak, and reflecting an advanced concern with the genetic viability of zoo populations of species and organisational mechanisms for implementing co-operative breeding schemes. Zoo professionals Tom Foose and Jonathan Ballou explained how zoos ought to function:

> It appears that the survival of many species, especially larger vertebrates, will depend upon assistance from captive propagation over the next century or more. Zoos and aquaria can and must serve as arks for vanishing wildlife. However, captive propagation can truly assist conservation of endangered species only if zoo and aquarium populations are managed genetically and demographically in a manner to reinforce, not replace wild populations. In the future, conservation strategies will ideally incorporate both captive and wild populations that are inter-actively managed for mutual support, that is, through regulated interchange of animals or at least genetic and demographic material that can be infused periodically into remnant wild populations or re-established in vacant wildlands . . . Reciprocally, the wild populations, even if only remnants, will still be subjected to natural selection and thus maintain some semblance of the characteristic genetic makeup of the species in the wild.[46]

The complexity of interchanging animals among institutions and between the captive and wild situations gave rise to the development of 'multi-institutional propagation programs' that would manage the myriad genetic, demographic and organisational considerations in co-ordinated breeding schemes for endangered species. Zoos in North America, Europe, Australia, New Zealand and Japan developed these Species Survival Plans (SSPs) under the auspices of their respective regional zoological associations. The Captive Breeding Specialist Group (CBSG), a specialist group of the IUCN's Species Survival Commission, was formed in 1979 to provide assistance and technical advice and facilitate co-ordination for captive breeding programs in the international zoo community. This organisation, based at Minnesota Zoo in the United States, has become one of the most influential forces in international, regional and national zoo conservation policy. In addition to these organising efforts, artificial and assisted reproduction techniques were increasingly being borrowed from agriculture and human medicine and applied to zoo populations of animals.

In the last two decades, zoo conservation policy development has been evolving in response to various criticisms and changing contexts, values and demands. These situations have enhanced and restricted the capacity of zoos to develop truly progressive conservation programs that reflect contemporary ecological principles. Essentially, the international zoo community has been engaged in continuing internal and external struggles to define and actualise a conservation identity for itself.

Although zoos had been facing problems such as providing humane treatment for their captive animals and the need to restrict their intake of wild animals for their collections, animal welfare and conservation issues were largely separate. It has been argued that the momentum for conserving endangered species has come, not from a concern for individual animals, but from ecological and utilitarian reasons.[47] Yet there were several organisations (Zoo Check, People's Trust for Endangered Species and International Fund for Animal Welfare) and several notable campaigns that drew heavily upon conserving species for compassionate, rights-based rationales (whaling, fur seals). Indeed, the late 1970s and early 1980s faced zoos with the growing importance of animal welfare issues within the conservation movement.

These concerns were embodied in the literature and in the formation of key associations. Jordan and Ormrod's book *The Last*

Great Wild Beast Show was published in 1978. The authors were not just concerned about the inadequate housing of many animals in some zoos. They wanted to know more about the quality of conservation programs in zoos. They posed some difficult questions that remain central to the contemporary zoo debate: is the purpose of zoos to entertain, educate, or study and preserve rare species? By the middle of the 1980s the Royal Society for the Prevention of Cruelty to Animals and a new organisation, Zoo Check, had joined forces to criticise the treatment of animals involved in zoo-based *ex-situ* conservation programs and stressed the need for conservation in the wild.[48] Today, both organisations continue their critique of zoo practices.

These matters pose a complex challenge for zoos, whose contemporary justifications for their practices emphasise the benefits that zoo-based programs have for conserving biodiversity. Zoos have defended some practices (such as killing individual animals that are considered surplus to the captive population of an endangered species) on the basis that the benefits conferred to species by virtue of *ex-situ* conservation measures far outweigh any harm or discomfort that may be imposed on individual animals. Yet, as we will see later, in addition to the problems that animal welfare interests raised, more sophisticated knowledge of conservation biology has also shed light on the very real theoretical and practical limitations of *ex-situ* conservation as a means for preserving biodiversity. Several international and national conservation policy statements and strategies have specified the auxiliary role that *ex-situ* methods should play relative to *in-situ* methods.

In response to these matters the international zoo community has endeavoured both to clarify and strengthen its conservation identity to itself and the general public. Several prominent zoo organisations have changed their names—omitting the word 'zoo' altogether—in an effort to convey an image that more accurately reflects their conservation mission. For example, the New York Zoological Society changed its name in 1993 to the NYZS—The Wildlife Conservation Society. The Society also altered the names of its five properties: the Bronx Zoo became the International Wildlife Conservation Park; the New York Aquarium is now the

Aquarium for Wildlife Conservation; Flushing Meadow Zoo was changed to Queens Wildlife Conservation Center; Central Park Zoo would now be known as Central Park Wildlife Conservation Center; and Prospect Park Zoo became Prospect Park Wildlife Conservation Center. Similarly, the *Captive* Breeding Specialist Group (CBSG) changed its name in 1994 to the *Conservation* Breeding Specialist Group.

Another very notable formal response to changing times was the World Zoo Conservation Strategy's release in 1993 at the IUDZG's annual meeting. The Strategy remains a central focus for zoo policy today. It has endeavoured to elucidate the *international conservation* roles of zoos and provides guidelines for implementing conservation-based policies, which will enhance the collective potential of zoos for protecting endangered wildlife. The document targets national and international policy—and decision-makers and local government authorities; government bodies, councils and benefactors of zoos and aquaria; zoo and aquaria professionals; and other conservation organisations, particularly government wildlife agencies. The initial release and subsequent use of the Strategy is indicative of the aspirations of factions within the zoo community that are eager to clarify zoos' international conservation role. Moreover, the Strategy symbolises an attempt by elites to rally the rest of the international zoo community in the struggle to stave off impending threats to biodiversity and perhaps threats to the zoo community itself.

Conferences and symposiums held during the 1990s reflect the zoo community's ongoing concerns with determining the boundaries of their conservation imperatives and their ethical obligations to non-human nature. These meetings also illustrate how zoo professionals negotiate among themselves for the acceptance and eventual adoption of what they believe to be appropriate policies at regional and organisational levels. In the United States a symposium was held in Atlanta, Georgia in 1992 to debate the future of zoos and aquaria, the ethical treatment of animals in captivity, and appropriate foci for conservation. In their endeavours to achieve as much consensus as possible, symposium participants were able to compile a list of conflict-reducing policy recommendations to the

American Zoo Association and to directors of captive breeding programs. The Australasian zoo community looked at similar issues at its annual meeting at Healesville, Victoria in 1996. The objective of this conference was to consider whether zoos were evolving in line with contemporary community expectations and zoos' conservation aims, or whether zoos were in danger of becoming irrelevant institutions, doomed to eventual extinction. While conference participants did not necessarily achieve consensus on all matters, they were able to openly articulate points of debate and clarify issues that needed consideration in their respective organisations and missions.

A particular perspective remains at the forefront of the debate and continues to influence policy choices: if zoological gardens and wildlife parks are to assist the restoration of biodiversity, they need to further refine their management skills of captive breeding *and* play a greater role in the area of *in-situ*. For example, the North American zoo community believes that if it is to have more of an impact on conservation its SSPs would have to strike a better balance between *ex-situ* and *in-situ* conservation efforts. As early as 1993, a more 'organised' effort was deemed necessary and an In Situ Conservation Committee within the AAZPA was formed to promote and facilitate field conservation initiatives by member institutions.

While the Australasian region does not have an *in-situ* committee *per se*, the need for zoos to engage in *in-situ* conservation programs is extensively endorsed. Several zoos vigorously pursue further involvement in inter-agency endangered species programs. There have also been calls for improving the integration between *ex-situ* and *in-situ* conservation programs and creating better working relationships with government wildlife agencies. Moreover, some members of the zoo community were, and remain, keenly aware of the need for more humility when forging relationships with outside agencies. They have cautioned against attacking criticism with grandiose statements about zoos' accomplishments, particularly in their relations with government wildlife agencies. Graham Mitchell, former Director of the Royal Melbourne Zoological Gardens, typified that sentiment:

> The key to success lies in integration with the activities of those legislatively charged with wildlife management at a local level, and with international conservation programs. In their publicity and in highlighting captive breeding achievements, zoos must re-inforce the notion of coordinated national and international collaboration (at both technical and political levels) as the essential ingredient and that the essence of conservation is in maintaining and securing habitat. They must never give the impression that they are going it alone or that they are even the key players.[49]

Other major discussions about policy priorities have included how important education is relative to zoos' other imperatives. While there are many in the zoo community who appreciate the importance of *ex-situ* programs linked to *in-situ* work for conserving endangered species, they frown upon the extent to which zoos' educational mandate is predicated upon rhetoric, and suspect that zoos' organisational (and financial) commitment to education are ineffectual. Recent organisational and funding adjustments to zoo education infrastructure in some organisations have further entrenched that skepticism. Other policy debates to emerge since the late 1990s in the Australasian region include the following.

- to what degree, and on what issues, should zoos be public commentators on environmental or conservation issues (they should avoid being 'political');
- how to balance 'for profit' activities (being commercially viable) with conservation, education, and animal welfare imperatives;
- how to develop effectiveness measures for conservation and education programs;
- determining parameters for the ARAZPA as a voice for the zoo industry;
- how to identify and deal with zoo adversaries.

The current political and administrative context of these zoo discourses is significant. Since the mid-1980s economic rationalism has become a powerful ideology and method of governance in many Western nations, including Australia. Public sector reforms now mirror business principles of the private sector. Consequently, zoos in both sectors operate largely according to a corporate management framework that calls for a quantifiable type of accountability. The realisation of ecologically-oriented conservation principles and programs in zoos is frustrated by this corporatised outlook which prioritises order and control over flexibility and participative processes, and rationalises the feasibility of most activities strictly in terms of economic efficiencies. This trend may represent the most recent phase in zoo evolution not accounted for by George Rabb's vision;[50] a period characterised by *corporatised* conservation.

The Future Ark

Zoos' conservation profile has been developed in recent times. This transformation has been shaped by shifts in human values of and attitudes towards non-human nature and by prevailing economic and political situations. Since the times of ancient menageries and later zoos, the practice of maintaining collections of wild animals conferred instrumental value onto animals, using them as indicators of power and wealth and sources of entertainment and amusement for élites. Many of these principles still inform contemporary zoos. A significant shift in zoo evolution was the spread around the world of a European model of zoos during the eighteenth and nineteenth centuries. Zoos became sites for pursuing scientific and educative endeavours (which would allegedly benefit the community at large) and were visited by members of the general public. By the middle of the nineteenth century zoos of this ilk were proliferating rapidly. Although now accessible to the general public, zoo development still tended to be instigated and sustained by groups of powerful and wealthy individuals. By the twentieth century a concomitant interest in conserving the integrity of the biosphere and exhibiting compassion towards sentient beings presented zoo professionals with substantial philosophical and practical challenges, the likes of which they had yet to encounter. The contemporary conservation role of zoos has emerged from these conditions.

What can we expect from zoos in the future? That will depend on how the zoo community responds to contemporary political, social, environmental and economic contexts. Will that evolution see a total transformation of the zoo as 'natural history cabinet' to the zoo as 'environmental resource centre'? Many leading zoo professionals will continue to try to guide the Ark on this evolutionary course. Their successful navigation will depend to a large degree on how they address difficult decisions about balancing multiple and sometimes competing imperatives. One major policy adjustment has been upgrading species management programs, which play a central role in zoos' stated contribution to biological conservation.

2

Passenger Lists and Procedures

And of every living thing of all flesh, two of every sort shalt thou bring into the ark, to keep them alive with thee; they shall be male and female . . . Of fowls after their kind, and of cattle after their kind, of every creeping thing of the earth.

Genesis 6: 19, 20 (King James version)

There are important differences between the story of Noah's Ark and the contemporary role of the zoo. There are, in effect, two different Ark policies. The biblical Ark embodies the expectation of the survival of life on Earth. Noah was able to fill his Ark with a representative sample of *every living thing*, from the largest of mammals to the smallest of 'creeping' beings. He was able to ensure their survival while the Ark was under sail. Once the floodwaters had receded, Noah was able to return those animals to their natural habitats where they multiplied, eventually re-populating the Earth. The reality of the modern Ark differs substantially from this biblical tale. It is predicated upon the combined efforts of an international zoo community which 'thinks globally and acts locally'.

Today's zoo professionals face many challenges that Noah was fortunate enough not to have to consider. The passenger list of the modern Ark is much shorter and much less representative of biological diversity than was the biblical Ark's, although neither Ark is known for its capacity to preserve the ecological relationships among its animal passengers and their habitats. A shortage of space on the modern Ark means that zoos provide passage for only a small fraction of the world's species. In addition, the modern Ark's passenger list privileges some types of animals: this includes the management and breeding of endangered species in captivity for eventual reintroduction to the wild. Given certain conditions on and the amount of time spent in the Ark, zoo professionals have had difficulty in breeding their animal passengers for successful reintroduction. Moreover, unlike the biblical floodwaters, contemporary habitat destruction continues largely unabated in numerous places. These trends make it very dangerous—near impossible—to lower the Ark's ramp to enable many of the 'passengers' to disembark. These difficulties are perhaps further complicated by the presence of several 'Noahs'—many of whom have varied notions of whom the Ark should carry and what course it should take.

Conserving Biological Diversity on the Ark

In general terms, 'conservation' represents both an ideology and a method for maintaining some semblance of the 'integrity' of the planet's biodiversity by safeguarding a state of non-human nature as we find it, as well as preventing wasteful or damaging practices. Concern for the decline of *species* indirectly represents recognition of the growing negative impact that human activities are having on natural habitats and the overall degradation of functioning ecosystems that support them. Zoo efforts to conserve endangered species become part of an overall effort to sustain biodiversity.

Biological diversity is the variability of all forms of living organisms and includes the terrestrial, marine and aquatic ecosystems of which these forms of life are a part and the ecological relationships that link them together.[1] There are several elements that are required for sustaining optimal levels of biodiversity. They can be likened to three separate decks on the Ark: on deck one is genetic diversity; on deck two is species diversity; and on deck three are communities. There must be sufficient levels of genetic diversity for species to maintain their reproductive vitality, resistance to disease, and capability to resist disease. A diversity of species allows for a range of evolutionary and ecological adaptations to particular environments. Finally, community-level diversity demonstrates collective responses of species to different environmental conditions.[2] Ecosystems that have greater levels of biological diversity will be more resilient. In the face of stresses such as human-induced habitat degradation, the system will have access to more options and be better able to revive itself in the face of these pressures.

One of the major tools that zoos use to conserve biological diversity is *ex-situ* conservation. When a species is threatened by extinction, and the threats to that species cannot be immediately removed, then individual animals from a species' community/population can be removed from the wild to be maintained in artificial conditions under human supervision. *Ex-situ* conservation is, in effect, the Ark. Government wildlife agencies, game farms, zoos, aquaria, botanic gardens, arboreta, and seed banks use *ex-situ* conservation. Individuals from *ex-situ* populations can be periodically released into the wild to maintain numbers and genetic variability in natural populations. In addition, research on these populations can provide further insights into a species' biology and methods for its conservation. Self-maintaining *ex-situ* populations can also reduce the need to collect individuals from the wild for display and research purposes, and can be used to educate the public about the need to preserve biological diversity.[3]

Through the use and promotion of *ex-situ* conservation as a tool to assist in restoring species and biodiversity, 'the Ark' has become a popular metaphor for both zoos' and other organisations' efforts to 'rescue' non-human nature from the threat of extinction. *Ex-situ* conservation became predominant in zoos during the middle of the twentieth century when many zoo professionals realised that zoos' consumptive practices of the past were no longer feasible. Rather than causing a drain on the shrinking supply of wild animals, zoo stocking policies would now have to make a positive contribution to endangered species conservation. It became critically important for captive breeding programs—irrespective of whether they were being conducted for the purposes of education, exhibitry or endangered species conservation—to reflect principles of conservation and population biology. The Ark could no longer be a thoughtless 'consumer' of wildlife, it had to maintain *and* 'produce' it by careful, selective practices.

Propagating and sustaining Ark passengers is no easy task. When animal populations are maintained in zoos over a period of time, their small size and subdivisions make them more susceptible to a loss of genetic variability, causing genetic degeneration and domestication. These conditions detract from the educational and conservation values of zoo animals because the species have a diminished capacity to reinforce or re-establish natural populations, and are unlikely to survive for several generations in zoos. Consequently, zoo professionals now use intricate, co-ordinated breeding programs that seek to maximise or at least maintain the genetic diversity of zoo populations of species and avoid in-breeding. These plans are fundamentally predicated on the pooling of zoos' efforts, as no single institution could provide the genetic or organisational resources that are otherwise needed to fill the Ark with animal passengers and then to sustain those passengers. In the short time that Noah's Ark was afloat, genetic diversity was not a problem.

Organising the Ark and Its Fleet

The Ark metaphor, as applied to the zoo, might connote the image of a single vessel into which we put many species. The reality, however, is that the Ark's total berths are distributed across hundreds of zoos around the world. These zoos act like a type of support vessel to the notion of the main Ark, and collectively, their disparate efforts actualise the Ark's mission. Other organisations—government wildlife agencies—can be considered part of the 'fleet.' A high degree of co-ordination is required to ensure that these vessels stay on course if a maximum number of species are to be 'saved'. The zoo community relies on several federative organisations to marshal the collective potential of the Ark 'fleet', to co-ordinate and control zoo policies and programs.

The World Zoo Organisation acts as the umbrella organisation for the international zoo community and operates under fairly formal procedures. Memberships are held by individual zoos and by national and regional associations. Regional zoo associations exist in North America, Europe, Australasia, Southeast Asia, Central America and Africa. The most highly developed and influential regional organisations are located in North America (the American Zoo and Aquarium Association—AZA), Europe (the European Association of Zoos and Aquaria) and Australasia (the Australasian Regional Association of Zoological Parks and Aquaria—ARAZPA). Approximately twenty-five countries have national zoo associations, which vary in size and the degree to which they influence and co-operate with international policy mandates. In addition to these geographic networks, there are several discipline-based associations that focus on educational, veterinary and collection planning issues. Two of the most significant organisations are the International Species Information System and the Conservation Breeding Specialist Group (CBSG). The International Species Information System co-ordinates the information needed for a global database on captive animals and is made up of a network of zoos and related institutions. The International Species Information System also houses the World Zoo Organisation secretariat. The CBSG, one of the largest of the IUCN's Species Survival Commission specialist groups, acts as global adviser and facilitator for regional *ex-situ* conservation projects and provides technical support to zoos for their captive breeding programs. This group functions less formally, acting as a sort of think-tank for zoo professionals from which ideas are then passed along to the World Zoo Organisation.

Regional associations are critically important to the smooth sailing of the Ark fleet. They provide administrative and professional support to zoos in their respective areas for managing species. They also oversee regional breeding programs, which are linked to and integrated with the global endangered species conservation programs of the CBSG. The AZA, which has 184 member-zoos, oversees the Species Survival Plan (SSP). There are currently eighty-four SSPs, which cover 136 individual species, a small proportion of the world's threatened species. The AZA's Wildlife Conservation and Management Committee approve all SSPs. Similarly, the ARAZPA governs the region's breeding scheme through the Australasian Species Management Program (ASMP). An ASMP Board of Directors, appointed by the ARAZPA executive, oversees the operations of the Program. In Australasia, thirty-two public and private zoos, aquaria and wildlife sanctuaries are members of

ARAZPA and participate in the region's management scheme. Table 1 shows eleven of the major ARAZPA-member zoos and the States (and Territory) where they are located.

Table 1 Australia's major zoos

New South Wales	• Taronga Zoo • Western Plains Zoo
Northern Territory	• Territory Wildlife Park • Alice Springs Desert Park
Queensland	• Currumbin Sanctuary
South Australia	• Adelaide Zoo • Monarto Zoological Park
Victoria	• Melbourne Zoo • Healesville Sanctuary • Victoria's Open Range Zoo
Western Australia	• Perth Zoo

Irrespective of their region, these breeding programs have highly similar regulations for how the Ark's animal passengers should be bred. These planning exercises can include members of the global zoo network, regional zoos, overseas organisations, non-government conservation organisations and government wildlife authorities. A species co-ordinator and a studbook keeper (sometimes the same person) collect, compile and maintain data about a particular species for a studbook. Together with representatives from other zoos who form a species committee, the population data for that species is analysed, and a breeding policy is formulated. These plans consider genetic matters, such as inbreeding and conserving genetic variability, and demographic issues, and offer guidelines for proper feeding, exhibit designs and other husbandry concerns. Zoos participating in the plan must agree to adhere to the mandates of the plan as closely as possible.

Unlike Noah, who had both a mandate and the room to take a sample of *all* the Earth's living creatures, zoo professionals face a serious lack of space in their vessels. Consequently, they must use regional breeding plans to devise their passenger lists. These lists are based on a number of characteristics, which form a scoring system. This system focuses on a species' degree of endangerment; existing regional co-ordination; the presence of formal conservation programs for particular species; and the extent to which those species can promote conservation-related education, an increase of knowledge (conservation-related research), or a combination of the two.[4] Species are then weighted according to how well they meet those criteria. Members of the zoo community designed these schemes to encourage all zoo professionals to orient their collections towards the more threatened species which also may be the focus of highly co-ordinated inter-agency breeding programs.

In Australasia, zoo professionals have created four categories of species for their species management plan—the ASMP. Category One contains many highly endangered species, such as the Striped Legless Lizard or the Golden Lion Tamarin, which may be part of an inter-agency conservation program. Not all species that are the focus of inter-agency breeding programs will be listed under this Category, although the aim is to achieve such a listing eventually. An ARAZPA-member zoo may not yet hold some species; a captive population may play an 'advocacy' role, rather than directly supporting management in the wild; or a program may still be under development. Highest priority is given to these 'first class passengers. Category Two species often involve some of the higher profile exotic species frequently associated with the more public profile of zoos such as the cheetah, cotton-top tamarin, scimitar oryx, Rothschild's giraffe, and black rhinoceros. Regional management is high for these species and mostly the efforts of the zoo community rather than outside agencies. Category Two species are the second priority and could be likened to business-class passengers. Category Three species are often the more charismatic mega-fauna which characterise zoo collections (for example, Sumatran tiger, orangutans, Prezwalski's horse, and Golden Lion Tamarin). They receive less regional management, but are being tracked and recorded with a view twards elevating their status to Category Two. They are not, however, being 'managed' more intensively as there are often prohibitive costs and/or challenging logistics to acquiring them, a process which slows their transition to a higher classification. Category

Four includes typically less threatened species that comprise more traditional portion of zoo collections (such as galah, Hamadryas baboon, lion), and are held by several zoos. This category can also include rarer species—which can be unique to a state region—and held only by one institution (for example, yellow-footed rock wallaby, long-beaked echidna). These species typically receive the lowest level of regional management. Category Three and Four species have the lowest priority: economy class.

Once Ark passengers have been assigned a classification, a plan is formulated for how they will be managed during their passage. A Species Management Plan (SMP) is based on a single species and targets those zoos holding that particular species. The plans consider issues of limited space and the implications that low genetic variability of that species have for its viability for long-term survival. The SMP accounts for the species conservation status as covered by other relevant programs assessing its predicament in the wild and may overlap with such programs. Plan objectives are also drawn from previous and existing research in conservation measures and assessments, general biology, and the species' history in captivity. Captive populations will be reviewed on the basis of their demographic and genetic characteristics, provenance, and population parameters and projections. Plan objectives will specify program duration, an optimal level of genetic variability to be targetted, and population parameters. Captive management strategies and recommendations are made to each institution carrying the species in question. Research priorities and plans are identified, as are schedules for implementation.

There are other mechanisms used to achieve consistent procedures for procuring passengers and for how they will be treated once on board the Ark. Taxon Advisory Groups (TAGs) review species management and/or collection planning recommendations (SMPs) for specific *groups of species*. These groups are typically based on an order or family, and will include species that are native and exotic to a particular region or country. In Australasia, there are TAGs for: reptiles, New Zealand birds, Australian non-passerines, Australian passerines, exotic birds, monotremes and marsupials, primates, carnivores, perissodactyls and proboscides, artiodactyls, and small placentals. All those species currently held by ASMP-participating institutions are covered by such schemes.

The international zoo community has invested considerable energy designing and implementing its policies for the Ark. How effective has it been in achieving globally and regionally-co-ordinated species management and captive breeding programs, which are meant to help restore threatened species to their wild habitats?

Patching a Sinking Ark?

The title of W. William Meeks' book, *Beyond the Ark*, embodies a discourse about the inadequacies of the traditional conservation methodology associated with the Ark metaphor.[5] The scientific literature has questioned the capacity of any *ex-situ* conservation program to contribute to the conservation of biological diversity, and numerous international conservation policies have designated these programs as an activity that should be secondary to *in-situ* conservation. Some of the problems associated with *ex-situ* conservation are specific to zoos' captive breeding programs. The nature and frequency of the zoo community's response to these challenges provides some insight into the future of the role of zoos in conservation.

It is has been well established that *in-situ* or *on site* conservation is the better strategy for conserving biodiversity. In Noah's story and in contemporary times, a 'flood' of a massive scale threatens the natural habitats of a multitude of living creatures. Unlike Noah's tactic of removing animals from their habitats in the face of danger, *in-situ* conservation seeks to manage the flood by maintaining secure habitats in order to ensure species' survival. This approach involves maintaining natural communities and populations *in the wild*. Populations are then able to continually adapt through natural evolutionary processes. In essence, species are more readily preserved in natural, or nearly natural habitats than in artificial ones such as zoos or botanical gardens.

Arguments against the Ark strategy which is typically brought in as an emergency measure—when a species has declined to such a vulnerable state that removing it from the wild to breed it in captivity is thought to be the only way to 'save' it—parallels promoting preventative medicine over traditional medicine. The longer we rely on emergency measures to address health care problems, the greater the incidence of disease and ill health, and the greater costs become. Many criticise captive breeding because they believe such 'last-ditch efforts' to save species are ultimately ineffective. Symptoms rather than the causes of problems are being addressed. Even where the best and most comprehensive scientific knowledge is made available for these projects, political and administrative obstacles can still confound most efforts.

A third major concern with the Ark methodology is its species-based approach to conserving biological diversity. While Noah's strategy of two of every species is comprehensive, it fails to address the need to protect all *three* levels of biological diversity. Meeks and many others believe we now need to use habitat, landscape, or ecosystem approaches to conserving biological diversity.[6] Unfortunately, abandoning an Ark mentality in favour of implementing ecosystem approaches to conservation necessarily involves a higher degree of biological complexity, scientific uncertainty and expanded political jurisdictions, and an increased potential for conflict and degree of difficulty in planning and management.[7] Moreover, despite widespread acknowledgment that species-based approaches alone are insufficient, there remain formidable gaps in both our scientific knowledge of ecology and in our practical understandings of how to incorporate more progressive ecological concepts into existing conservation infrastructures.[8]

Like other conservation schemes, zoo-based captive breeding programs are plagued with a series of technological, policy, organisational and symbolic complications. As a result, professionals inside and outside the zoo community and conservation advocates are concerned about relying on the Ark strategy that underpins zoos' species management and breeding programs. Collectively, these problems constrain the Ark's (the international zoo community's) potential to support the conservation of biodiversity.

By the late 1980s and throughout the 1990s, considerable attention was (and continues to be) paid to the global zoo community's poor performance in establishing self-sustaining, viable populations of animals and its limited success breeding either vertebrate or invertebrate species in captivity. In 1993 one team of conservation researchers compiled some sobering statistics showing that, of the 4200 species in captivity, only 1200 had been bred successfully.[9] Essentially, the physiological, psychological and environmental needs of the Ark's passengers have not been catered for. There are also significant problems with breeding animals that are too closely related. Many zoo populations of animals were established before knowledge of the principles of conservation and population biology was very advanced. In 1992 one analysis of zoos' captive breeding programs found that only nine threatened mammalian taxa had captive populations over the recommended 500 specimens, and a further fourteen had captive populations exceeding 250.[10] Years of breeding closely related animals has resulted in a substantial loss of diversity which in turn still threatens the long-term viability of zoo populations.

The ultimate success of the Ark mission hinges upon successful reintroductions of endangered species to their natural habitats. Zoos have had some notable successes. An international survey of reintroduction projects showed that 59 per cent of those efforts drew on captive populations held by zoos.[11] Some successful reintroduction programs that have relied on zoo-bred animals includes the wood bison, plains bison, Arabian oryx, Alpine ibex, bald eagle, Harris' hawk, peregrine falcon, Aleutian goose, bean goose, lesser-white fronted goose, wood duck, masked bobwhite quail, Galapagos iguana, pine snake and the Galapagos tortoise.

Despite these achievements, overall success rates for reintroduction programs have been low: history suggests there are problems with establishing wild populations from animals bred in captivity. The same survey that documented reintroduction successes, found that of the 145 reintroduction projects involving captive bred vertebrates, only sixteen of them were able to re-establish wild populations. Others have made similar findings. These poor success rates are caused by several problems that have detrimental impacts on a captive-bred animal's capacity to survive in the wild. A collective of conservation and zoo professionals conducting a review of captive breeding programs in most North American zoos found in 1996 that many released animals had significant problems with foraging for food, avoiding predators, or interacting properly with their own species.[12] Zoo professionals have gone to great pains to use their scientific and practical knowledge of wild animal behaviour to design exhibits that simulate wild conditions as closely as possible. However, it appears that many captive environments have not provided the conditions and stimuli necessary for some species to develop appropriate survival behaviours. These problems were found to apply to both learned and genetically acquired behaviours. In effect, some species have been domesticated during their passage on the Ark.

Outbreaks of disease on the Ark has limited the success of reintroduction. Despite substantial advances in zoo veterinary medicine since the middle of the twentieth century, disease among captive populations remains commonplace. An international exchange of animals and a diverse array of species in close proximity to one another, means that many animals have been exposed to exotic pathogens to which they have not yet developed resistances. In the review of captive breeding programs mentioned above, several basic preventative veterinary guidelines for avoiding disease outbreaks were offered. Yet the researchers found that for a majority of programs at that time, few of these precautionary steps were taken because of their exorbitant expense.

In addition to the costs of controlling disease, general costs challenge the sensibility of using *ex-situ* conservation as a broadscale approach to conserving biodiversity. The average cost for endangered species recovery efforts in the late 1990s was estimated to be US$500 000 per year per species.[13] And in 1994 the Golden Lion Tamarin project, which some members of the zoo community heralded as a spectacular success story, cost approximately US$7.5 million.[14] The Zoological Parks Board of New South Wales received a State Government grant of $50 000 in 1992/93, and considerably more than that in corporate sponsorship in the latter part of the decade, to help fund its participation in the international captive breeding program for black rhinos.[15]

The comparative costs of some *in-situ* programs highlights how expensive it has been to maintain the Ark. In their 1994 inquiry into zoo practices, the World Society for the Protection of

Animals and the Born Free Foundation estimated that maintaining a single black rhino in captivity cost approximately $16 800 compared to the $1000 it cost to protect its wild habitat.[16] The cost of keeping African elephants and black rhinos in zoos in the early 1990s was said to be fifty times that of protecting equivalent numbers in the wild in Zambian National Parks where one square kilometre of park was patrolled for US$400 per annum.[17] Maintaining the whole Serengeti ecosystem was estimated to cost US$500 000 per year, a price comparable to that of sustaining a viable population of five species of primates in North American zoos.[18] One study found the cost of a species conservation program increases from tenfold to ten thousandfold at each of the three levels of intervention: species managed where they occur naturally, intensive onsite management, and species restricted to or heavily dependent upon zoos.[19]

All economic considerations aside, the Ark's ability to conserve species could be questioned on the basis of how representative of threatened species its passenger lists have been. The World Conservation Monitoring Centre conducted an extensive survey of *ex-situ* conservation programs in the early 1990s.[20] This survey found that of the 629 mammalian species threatened on a global scale, 20 628 specimens from 140 threatened species were held in zoos. It also found that of the 15 per cent of the world's species under threat, the proportion of them represented in captivity was 22 per cent. Only 10 per cent of the global capacity of approximately 200 000 mammal specimens consist of threatened mammal taxa. By the middle of the 1990s, another study compiled similarly sobering statistics. Taxa in zoos and reintroduction represented a mere 1 per cent of animal species found on earth; stretched to their full capacity, all the zoos in the world could sustain at best only 900 species by the year 2000; and only sixty-four species management programs were established in the United States.[21] As early as 1986, members of the zoo community had calculated that there was only enough space in zoos for a mere 1000 taxa.

The limits of space and resources available to the modern Ark continues to present very real challenges. So too does the expansiveness and complexity of the problem of biodiversity loss. Given these conditions, the need to conduct existing programs as effectively as possible and to capitalise on regional and global co-operation among the different zoos making up the Ark fleet becomes ever more critical. Why are there not more threatened species represented in zoo collections? How consistently are zoos implementing regional collection planning and species management priorities today? The answer lies in part with variations in organisational preferences and local and national settings.

Setting Course in Varied Environments

The success of the Ark's conservation mission depends on how well the fleet—made up in part by hundreds of zoos around the world—is co-ordinated. In order to maximise the use of limited ecological and organisational resources, zoo professionals must consistently adhere to global zoo standards and regional collection planning measures. Several zoos are typically involved in managing a single species which often entails exchanging animals, encouraging or discouraging the breeding of certain genetic lines, or expanding or reducing a total population size. These arrangements are based on voluntary participation by zoos. Hence, an unprecedented level of national and international co-operation and effective communication is required to co-ordinate the activities of hundreds of organisations.

Effective co-ordination of the Ark's fleet is one of the most central policy issues for zoo professionals. It is essentially about when to honour institutional sovereignty over regional priorities. The World Zoo Conservation Strategy stresses uniformity in zoo principles and programs in order to achieve conservation goals:

> No matter how well organised, how well led and how well trained zoo personnel are, a major contribution to conservation can only be achieved if zoos work together and exchange information. No zoo has all of the necessary knowledge and experience within the limits of its own confines. Each zoo is dependent on others for information in a variety of specialised areas. Zoos have made much progress in the sharing of knowledge and experience with others: yet zoos should still strive for even more intensive cooperation.[22]

Despite calls for increased co-ordination, individual zoos may choose *not* to follow the recommendations of global or regional plans. In an international planning workshop organised by the World Zoo Organisation in 1998, participants identified a lack of commitment and co-operation among some members of the zoo community as a major constraint to integrated and co-ordinated zoo programs.[23] For example, Table 2 shows that, between 1996–1998, the highest priority was placed on Australasian zoos holding Category One species, yet these species represented a small proportion of the region's overall collection of which 92 per cent consisted of the lowest priority species—Category Four.

Ironically, the modern Ark's passenger lists have typically favoured 'economy class' passengers, the charismatic mega-fauna which are now *lower* priorities in regional breeding schemes. It has been hard to break this tradition, confounding goals of regional consistency. In 1995, the ASMP's Census and Plan identified that the Reptile/Amphibian group of species was not sufficiently represented in the regions' zoos. By 1998 it was still an issue of concern, as the TAG Convenor for the Amphibian group noted:

> A relative lack of interest by the zoological and aquarium community in the display and husbandry of amphibia continues to severely limit the achievements of the TAG. To date there has been little directed effort with these taxa, little is known about care and successful breeding techniques for Australian frogs and very little information has been published or collated on the zoological industry's experience.[24]

It remains harder to convince zoos participating in regional planning exercises to feature less popular taxa. Collection management decisions appear to remain highly influenced by what people perceive to be the more effective power of 'charismatic mega-fauna' —the carnivores, rhinoceros, and other species—to draw in visitors, hence raising zoo revenues. Indeed, some of the exotic, larger vertebrate species can command a considerable level of public attention, and their commercial popularity may confound species management planning.

Table 2 Species held by ARAZPA-member* zoos

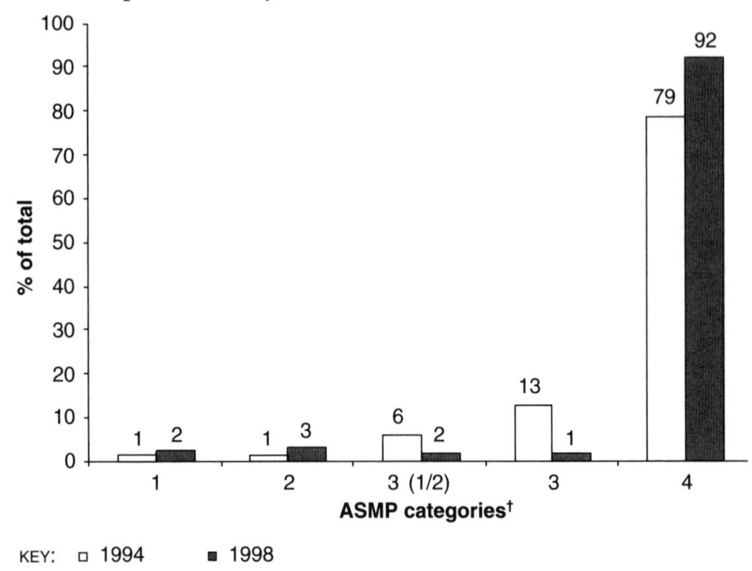

* *ARAZPA administers the ASMP which co-ordinates zoo collection plans within Australasia by providing administrative and professional support to the region's zoos for managing species and for integrating the region's programs with the global endangered species conservation programs. Zoos who are full institutional members of ARAZPA are included in ASMP and sign a declaration agreeing to follow ASMP policies as closely as possible.*

† *The ASMP uses four main categories to prioritise the species held by ARAZPA-member zoos. Category 1 is the highest priority, while Category 4 is the lowest. Category 3 (1/2) species are those which qualify for the higher categories (1 or 2), but are being held by a zoo for educational or 'advocacy' purposes or for whom particular programs have not yet been fully developed.*

SOURCE: Compiled from Australasian Species Management Plan, *Regional Census and Plan*, 1994 and 1998.

There is now a trend for zoo professionals from industrialised nations to provide institutional and financial assistance to 'third world' nations by helping them develop their own zoos, establish endangered species programs for which captive breeding may be an option and, in some cases, remove endangered animals from these countries to transfer them to captive breeding programs in European, North American, or Australasian zoos. These programs do help to produce more modern zoos or more individuals of a threatened species. However, their effectiveness in arresting the processes threatening those species in the first case is more dubious. The Sumatran rhino is a case in point. This highly endangered species has been the focus of intensive conservation efforts (involving government officials and several zoo-based international funding agencies and conservation agencies), which are oriented primarily towards its capture and attempted breeding. Allan Rabinowitz, a conservation professional who has been closely associated with the program, has criticised its practices.[25] He suggests that these expensive, ineffectual—but politically expedient—strategies have been used, rather than implementing *in-situ* techniques that target the last remaining wild populations of the species in Indonesia and Malaysia. In the case of planning for different species of rhinoceros in Australia and the United States, competition and conflict between institutions and among regions for holding the rhino has been high and posed threats to regional consistency. The 1994 ASMP noted that those zoos that had traditionally 'lead the way' with captive breeding programs for rhinoceros were compromising regional planning initiatives.

Behind the question of honoring local imperatives before regional priorities lies the very real need to sustain sufficient gate-takings to 'pay the bills', but, perhaps more importantly, the perception that people will only come to the zoo if there are large, exotic species featured there.

Who should be Noah?

Effective operation of the Ark depends on a well-co-ordinated fleet of vessels. Yet this fleet is not just made up of hundreds of zoos. There are a host of government authorities and non-government organisations that have an interest in and responsibilities for conserving threatened species. Issues of logistics aside, local, state and national political settings raise questions about who is (or ought to) be in command of the Ark. The way such contests are resolved is highly relevant, not just to the nature of the Ark mission but also to how well it is carried out.

The example of Australian federalism shows how a country's political structure creates, in effect, a system of organisation for the Ark's fleet. The chain of command in this system is highly complex and often ambiguous. In Australia, federalism is characterised by three levels of government, each possessing distinctive powers and responsibilities. This system of governance results in a complex network of relations between and among the different levels of government. These levels are made up of a multitude of agencies, organisations and individuals. Relations among these actors are changeable and subject to conflict, given ongoing struggles for power and differences in values. Where environmental matters are concerned, these settings are particularly relevant to forming effective and cohesive policies. Environmental problems, such as widespread species extinctions, are extremely complex and are not readily dealt with within the confines of any single authority or organisation.[26]

These challenges of making and administering all forms of environmental policy in federal states have been recognised.[27] There is now an increasing emphasis on better co-ordination among vessels of the Ark fleet. Many legislative, strategy and policy mechanisms have been created in an effort to foster more co-operative and co-ordinated programs among federal, state and local environmental authorities. Yet the growth of these programs has given rise to a proliferation of multiple points of access to decision-makers in all kinds of environmental decisions. Here varied priorities and the quality of working relationships become central to the success of achieving environmental protection.

In Australia, there is a convoluted network of Commonwealth and State policies and legislation that address endangered species conservation, wildlife protection, and animal welfare. All levels of government share responsibility to protect flora and fauna in their respective jurisdictions, to use their best endeavours to ensure survival of species and ecological communities, and to conserve areas critical to protection of such flora and fauna. Several administrative arrangements and legislative instruments are used at both Commonwealth and State levels to protect native and exotic species that are immediately or potentially threatened by population declines, as well as non-threatened native species. All wildlife species are afforded a special protective status that creates a mandate for their conservation, as well as restricting the ways in which they can be manipulated. For zoos, these mechanisms influence what passengers can be loaded on the Ark, and how they will be treated. Zoos' role in conservation is directly and indirectly influenced by regulations on the import and export of native and exotic animals which determine the degree to which zoos will be involved in endangered species conservation, and specifying minimum standards of care for captive animals.

Zoos communicate both as a single community—and as individual institutions—with Federal and State environment departments, and their respective wildlife units, about administrative and regulatory matters and joint endangered species programs. There are several authorities concerned with zoos' goal of regionalising and globalising their collection plans and expanding their involvement in endangered species conservation. These are depicted in Table 3. Two of those agencies are the federal government's Wildlife

Table 3 The Australian zoo community and government agencies

National Parks and Wildlife Service (NSW)	• Taronga Zoo • Western Plains Zoo
Parks and Wildlife Commission (NT)	• Territory Wildlife Park • Alice Springs Desert Park
Dept of Natural Resources (Qld)	• Currumbin Sanctuary
Environment Australia (C'wealth) *Threatened Species and Communities Unit* *Wildlife Protection Authority*	• **ARAZPA/ASZK** (the zoo community)
Dept of Environment and Heritage (SA)	• Adelaide Zoo • Monarto Zoological Park
Dept of Natural Resources and Environment (Vic)	• Melbourne Zoo • Healesville Sanctuary • Victoria's Open Range Zoo
Dept of Conservation and Land Management (WA)	• Perth Zoo

Protection Authority and the Threatened Species & Communities Unit (TSCU).

The Wildlife Protection Authority (WPA) regulates the export and import of threatened species as per Australia's obligations under the Convention on International Trade in Endangered Species (CITES). Given the emphasis on regional animal collection plans, the zoo community—and the Australian zoo community especially, given its relative isolation—is highly dependent on inter-zoo animal exchanges within and outside the region. The policies affecting zoos' ability to procure passengers—import exotic animals and export native species—are critical.

The policies of the TSCU and its State counterparts are central to Australian zoos' ability to fulfil their Ark imperative. Where native species are concerned, these agencies are higher up the chain of command in the Ark fleet than zoos. The TSCU liaises with State wildlife authorities to establish policy objectives and provides funding for the recovery of threatened native species and ecological communities. These measures have been provided for by the *Endangered Species Act 1992* (Commonwealth) (ES Act). In 1999 the *Environment Protection and Biodiversity Conservation Act 1999* (EPBC Act) replaced this Act, as well as several other pieces of environmental legislation. The current legislation, and its predecessor, have and will provide the overall context for endangered species protection at a federal level, and a vehicle for zoo involvement in *ex-situ* and *in-situ* conservation through the preparation of Recovery Plans for individual species of native animals. The ES Act was criticised on the basis that it suffered from weaknesses similar to those lodged against using Ark methods, using a species approach to conserving biological diversity. Other criticisms of the repealed ES Act were that that it lacked comprehensive powers to provide for co-ordinated, national endangered species management and failed to resolve who has jurisdiction over conservation programs.[28] Since the EPBC Act was passed, claims have been made that threatened species protection will be improved because the Commonwealth's powers to protect a host of biodiversity components have been expanded and the delineation of Federal and State governments' responsibilities has been made clearer.[29]

Nevertheless, while Recovery Plans can be thought of as providing a course for select vessels in the Ark fleet to follow, questions about who is commanding that fleet may still arise. Threatened species often occur in more than one State or Territory. In these cases, issues concerning sovereignty, logistics and duplication of efforts among Commonwealth and state agencies, as well as among the zoos may be more complicated. Recovery Plans tend to prescribe necessary actions throughout a species *range*, rather than in the single State or Territory in which that species occurs. According to the new EPBC Act, Recovery Plans may now be written by the Federal government for species found wholly within, partly outside or wholly outside a Commonwealth area, and these Plans may be made in consultation or jointly with the relevant State government(s). In these cases, a high level of co-operation and co-ordination is required among the Ark fleet—the different levels of government, the various state agencies and the zoos. Captive breeding may be included in a Recovery Plan. If so, the selection of a particular zoo (or zoos) will be based more on the expertise and resources that a zoo can make available rather than on whether a zoo is located in the same State as the wildlife agency conducting the program. Additionally, Recovery Plans based in one State can involve zoos in other states. Melbourne Zoo has assisted Western Australia's Department of Conservation and Land Management in a breeding plan for Pale-Bellied Frogs. Sometimes wildlife authorities will use several interstate zoos. The Victorian Department of Natural Resources and Environment's recovery effort for the Eastern Barred Bandicoot involves Healesville Sanctuary in Victoria and the Western Plains Zoo in New South Wales. In these cases, zoos are conducting relationships with the federal wildlife authorities and several State wildlife agencies at any given time.

There are other levels of interaction and manoeuvering among fleet vessels that should be considered. The responsibilities of the Commonwealth's TSCU do not preclude the States from preparing their own plans, although they may be somewhat tied to the Commonwealth by the need to seek funding for their efforts. State governments have varied in their approaches to protecting species. Some States have created more innovative and effective legislation and management approaches, others have made drastic cuts to their endangered species management units, such as the Victorian Liberal government's downsizing of its Department of Natural Resources and Environment. Given the zoo community's interest in being involved in inter-agency programs for restoring native species, they must be able to respond to the changing priorities of the federal and State governments. Navigating their way through this fragmented and dynamic policy landscape is no easy task for zoo professionals who are faced with their own maze of intra- and inter-zoo relations. In the 1996 ASMP Regional Census and Plan, Gary Slater made this astute observation:

> The reticence by Wildlife Agencies in adopting a multi-zoo approach to their species recovery objectives is understandable. The region has been slow in working together in order to develop viable collective approaches that wildlife agencies can consider. In order to ensure that the valuable resources of zoos are working effectively for priority programs, the collective approach to conservation programs will need to gain currency at executive management levels within our zoos, and to be promoted and supported by executive management within internal management strata of zoos.[30]

When fragmented policies inside and among zoos are set against a backdrop of the varied principles and practices of government wildlife authorities, the effect can be to scatter the Ark's fleet in different directions, ultimately limiting its collective capacity to assist endangered species conservation.

Perceptions of the Ark

Consistency in Ark policies is closely related to the quality of working relationships among the crews of the Ark's fleet. These exchanges are influenced in part by how people view the missions of their respective 'vessels'. Attitudes towards the role of zoos in conservation provide an example of how social dynamics might affect the nature of endangered species restoration efforts. For a considerable part of its history, the Western zoo operated in a domain that was largely separate from wildlife conservation and environmental policy-makers. As environmental concerns have spread to the zoo, zoo professionals increasingly interact with government wildlife agencies, non-government organisations (NGOs) and community groups in order to realise their conservation goals. In this 'community' of conservation professionals and advocates, there are divergent political philosophies and commitments concerning environmental issues and, more specifically, matters relating to endangered species conservation and zoos.

Discussion of passenger lists on the Ark has demonstrated that the zoo community values involvement in inter-agency recovery efforts for exotic and native species. Irrespective of their country of origin, not all Recovery Plans necessarily include *ex-situ* measures and, where they do, this still does not guarantee zoo participation. The Australian Federal government's Recovery Plan and Funding Proposal Guidelines specify that Recovery Teams 'may include . . . representation from captive breeding institutions if appropriate'.[31] In this case, recovery plan authors, be they from Commonwealth or State agencies or from the scientific community, determine the need for *ex-situ* conservation, and are likely to select what organisation will provide that service. This is a critical juncture in decision-making. Here individual perceptions, and the existence and nature of previous relationships with zoos, are critical determinants of whether zoos will be asked to participate in the recovery planning effort and of how those relations will proceed.

There is strong evidence to suggest that people have different ideas about who should be commanding the Ark fleet, while some question the very notion of the Ark, in so far as it is epitomised by zoos. I surveyed some members of the conservation community in Australia between 1993 and 1997 and found certain commonly held opinions.[32] When asked what they believed distinguished their own agencies from zoos, responses indicated that people felt the conservation scope of zoos has been narrower than that of a wildlife agency, and that zoos employ methods which do not sufficiently or appropriately address wildlife conservation. Some members of the zoo community might claim here that zoos' mission is not meant to be solely about restoring wildlife. The survey data suggests that, despite zoos' claim *not* to be aggrandising their role, members of the conservation community have perceived zoos as saying otherwise—that they *are* critically important conservation organisations. Some within conservation agencies thought that the traditional conservation 'territories' of government and conservation groups were being infringed upon by the relatively new goals of zoos. Most of the people from government wildlife agencies that were surveyed felt that the statutory responsibility of the government wildlife agencies should be guiding zoo involvement in wildlife conservation, certainly with respect to native species and perhaps less so regarding exotics.

Other views supported critiques of the Ark methodology made in the scientific literature. When discussing principles of conservation biology, respondents from the conservation community argued that captive breeding was costly, had a high rate of failure and detracted from the importance of conserving habitats. Other concerns of respondents included the genetic and behavioural degeneration of species held in zoos over extended periods of time. Some people were worried that the rigidly taxonomic concerns of zoos' species-based approaches addressed merely 'the result of the problem not the cause'.

There is no doubt that some members of the zoo community are aware of the fundamental conflicts between their institutions and certain other conservation agencies and missions. They have cited problems furthering their role in endangered species conservation due to their relationships with government wildlife authorities. These problems have typically been caused by difficult personal interactions, agency rivalry and suspicion, limited communications given bureaucratic restrictions, competition for funds, and a clash of organisational cultures and insufficient acknowledgment of zoos' expertise.

Disparate organisational and professional goals and overlapping policy domains are significant issues for productive inter-fleet relations. When looking to explain where and why some of these barriers have been crossed, one is likely to find the presence of long standing personal associations among key players in the respective organisations.[33] Friendships, and previously shared work experiences among zoo and wildlife agency staff, have enabled these people to gain greater knowledge about each other's workplaces, to de-bunk myths about either organisation and to learn about opportunities for joint programs. These informal links have been instrumental in creating and maintaining zoos' formal connections to the rest of the conservation community by ensuring that zoos are able to participate in endangered species recovery efforts or co-operative education programs.

It remains that some zoos will have better working relationships with their respective and other State wildlife authorities than others. In Victoria, animal management staff from the Melbourne Zoo, Healesville Sanctuary and threatened species managers and project officers from the Department of Natural Resources and Environment have worked together on several recovery programs. These people have a long history of shared professional and personal ideologies. Conversely, in Western Australia the relationships between the Department of Conservation and Land Management and the Perth Zoo were strained for some time. The Zoo identified the Department as a 'competitor' in the early 1990s in its Draft Business Strategy, and there were persistent negative attitudes towards zoos and poor relations between senior managers in the two organisations for several years. In Queensland prior to 1976, Currumbin Sanctuary had never been registered by the State's National Parks and Wildlife Service. While the Sanctuary's historic commercial popularity has facilitated the Wildife Service turning an official blind eye, relationships between the two organisations were supposedly improving due, in part, to the Sanctuary's affiliation with the National Trust.

Navigating the Shoals

The above might appear a rather sobering compilation of facts and figures relegating the Ark mission and methods to the realm of irrelevance for conserving biodiversity. This is only part of the whole picture. Important and encouraging policy developments have been made in the last decade that address some of the weaknesses of zoo-based conservation programs. Some of these efforts are small and subtle, while others exist on a larger scale and are more readily observed. What makes all of them significant is the willingness of some members of the Ark's fleet—be they from the zoo community or government wildlife authorities—to recognise the problems their institutions face and to search for new ways to redress those dilemmas.

Several different kinds of actions are being taken by a variety of players at regional, organisational and personal levels to strengthen the zoo community's ability to improve its *ex-situ* and *in-situ* programs. These actions function at a formal and informal level. First, there is an increased awareness among many zoo professionals of the ideological and practical differences in approaches to conservation between zoos, government wildlife authorities, and NGOs. Second, there are conservation professionals and activists who appreciate the changes zoos have undergone in recent times. Third, an increasing effort has been made to formalise the links between zoos and their respective wildlife authorities in order to standardise inter-agency procedures. Fourth, these arrangements have gone some way to achieving greater consistency among inter-agency goals and outcomes. In addition, they have provided numerous opportunities for informal interactions among zoo and wildlife agency staff which, in turn, has facilitated greater understanding and appreciation of organisational differences. Finally, some zoo professionals have been successful in designing and implementing innovative evaluation programs that target some of the persistent weaknesses in zoos' *ex-situ* conservation schemes.

Zoo professionals' commitment to improving the zoo's conservation performance appears to be closely linked to their willingness and ability to redress policy differences and conflicts between their institution and other conservation organisations. Remarks from interviews conducted between 1993 and 1997 with Australian zoo professionals who met regularly with wildlife authorities and NGOs, suggest that the zoo community has been partly responsible for propagating its own problems. One senior manager said, 'Zoos have muddied up the waters . . . [we] haven't helped ourselves by not clearly stating what our purpose is'. Along the same lines, another senior manager stated that since 'zoos and [wildlife agency] haven't defined the problem or program very well . . . the two agencies are seen to be doing the same thing, which will lead to conflict'. The situation was certainly not viewed as being hopeless. 'Zoos are resources for conservation', said one senior manager. He believed, however, that the zoo's role 'has to happen in conjunction with [government] conservation departments'. The solutions offered consistently pointed to the zoo community needing to acknowledge its auxiliary role in conservation, relative to the federal and State wildlife authorities' mandates, and to appreciate its responsibility to clarify any misunderstandings it has created. 'We should not be trying to take the place of the conservation role of state authorities', said a senior manager, 'We should be providing our experience to assist them.' One senior manager felt that zoos could improve their credibility 'without being threatening', if they could 'be seen to be the stable body in the background that complements conservation organisations'. Another senior manager believed that honest dialogues between zoo and conservation agencies were critical. She was interested in facilitating 'a much greater level of open discussion and debate about ideas and strategies than perhaps we get'.

In a similar spirit of co-operation and understanding, some members of the conservation community appreciated the wide scope of things zoos could do for conservation. When referring to the efforts of zoo professionals, one senior manager in an Australian government wildlife agency said, 'I prefer to believe that people's intentions are real . . . they are working very hard to make the World Zoo Conservation Strategy part of their charter . . . and we are very pleased to see that communication has improved . . . It's quite clear they are genuine.' Another senior manager said, 'our relationship with zoos is very good from an administrative point of view . . . ARAZPA has made a major step forward in fostering relations.'

Formal and Informal Means

Social dynamics are central to efficient Ark operations. Increasing this kind of understanding is critical for the success of zoos' conservation programs. Current global and regional species management plans reflect the belief held by a growing number of zoo professionals that zoos should improve the relevance of their endangered species breeding programs by ensuring that they are linked increasingly to *in-situ* conservation schemes. For zoos holding threatened species that are native to their respective countries, this means having high quality working relationships with their State or federal government conservation initiatives. In the case of maintaining threatened exotic species, zoos may be supporting foreign governments or NGOs.

Individual zoos and regional bodies have made important efforts to clarify and subsequently improve their relationships with government wildlife authorities. There are several examples where individual zoos and regional bodies have made an effort to develop closer, formal links with government authorities. In Australia, the Zoological Parks Board of Victoria (ZBV) signed a Memorandum of Agreement with the federal environment agency in 1994 in order to clarify the Zoo Board's policies and facilitate joint conservation programs between itself and this authority. This formal agreement resulted in part from a joint project between the Zoo Board and the federal government that was already being developed—the establishment of a biosphere reserve in South Australia. Signatories to the agreement agreed to pursue 'cooperative endeavours in nature conservation, ecologically sustainable development, and the conservation and management of biological diversity both in Australia and overseas'.

The ARAZPA has also used formal and informal means to establish closer working relationships with wildlife agencies. The Association appointed an Executive Director, Christine Hopkins, whose responsibilities have included acting as an industry representative at non-zoo fora and facilitating better communications between government agencies and zoos. This appointment has helped to address problems associated with a lack of cohesion among the region's zoos. Until early 2000 when she stepped down from her position, Ms Hopkins made a concerted effort to have staff from the TSCU and WPA, and from state wildlife agencies, attend zoo conferences and seminars and serve as representatives on zoo-based planning committees such as the TAGs. Similarly, zoo industry members were invited by the TSCU to attend the Endangered Species Recovery Process Conference held in December 1995. These kinds of interactions have gone some way to facilitating improved understanding about the respective organisations and generating shared outcomes.

Another recent ARAZPA initiative for improving co-ordination among zoo and wildlife agency staff was a workshop designed to review the current practices of Australasian species management programs. The significance of this workshop is that it demonstrates that many zoo professionals recognise the gaps between zoos' formal goals to actualise an Ark mission and their actual performance, and hope to find ways to address those discrepancies. Workshop participants essentially considered how the Ark might be filled with more first class passengers. That is, how might zoos direct *more* resources towards species recovery processes in general, and Category One species—the region's highest priority—in particular? Representatives attended the workshop from federal and state wildlife agencies and zoo staff from the Australian region who manage high priority species.

The workshop clarified why zoos should seek to reorient their animal collection plans and direct more resources to species recovery processes in Australasia. Participants noted that there is a clear, formal mandate from global zoo industry policy statements, such as the World Zoo Conservation Strategy, major national and State biodiversity strategies, and individual zoos' mission statements that this is what modern zoos ought to do. A fundamental part of those discussions was the conclusion that zoos have *not* yet met their full potential in assisting the conservation of biological diversity. Second, workshop participants determined that the existing process for species recovery was dependent on inter-agency partnerships and provided zoos with the opportunity to support local conservation issues and to contribute a range of skills. Participants then went on to identify a range of zoo resources which could be used to support recovery programs, such as small population management and captive husbandry expertise, a range of facilities, and access to fundraising and publicity mechanisms.

Co-ordination of zoos and wildlife agencies was another important focus of conversation at the ARAZPA workshop. Several aspects of zoos' structures, operations and cultures were identified as inhibiting more productive partnerships with wildlife authorities.

Historic and current practices deriving from commercial imperatives were noted as problems. These include a lack of commitment from 'business-minded CEOs' who devote too many resources to collecting and displaying exotic species in expensive exhibits. In addition, competition among zoos may foster collection planning akin to a 'stamp collecting' mentality. It may be easier to locate resources for these kinds of exercises than for funding the more obscure, but highly threatened species. These conflicting priorities contribute to what was felt to be a lack of specific policies designating what priority inter-agency conservation work should take.

Attention to vague policy prescriptions could be used to address another set of organisational problems. ARAZPA workshop participants identified a lack of appreciation by zoos for the need for sufficient and appropriate resources. In their enthusiasm to work on inter-agency recovery projects, some zoos may seek involvement knowing that another zoo may be more qualified to do so. In other instances, a zoo cannot afford to allocate staff to a project, or are unable to supply staff or financial resources in a timely fashion. Sometimes the zoo staff assigned to a recovery project may not be given sufficient authority to operate autonomously. These problems, as well as varying husbandry practices, have led some staff in the government wildlife authorities to view zoos with suspicion and inhibited the formation of links between them.

We saw that zoos are not the only 'vessels' in the Ark's fleet, and that they share responsibility for conserving wildlife with government wildlife authorities. In the spirit of achieving improved communication for the entire fleet, workshop participants also considered how wildlife agencies create obstacles to zoos being involved in recovery efforts. Their findings were consistent with the survey findings mentioned earlier. Some wildlife professionals have suspicious and negative attitudes about zoos which can cause them to overlook or trivialise the expertise and facilities available there. People may be at odds with the ideology of zoos and *ex-situ* conservation methods, with some starting to look increasingly towards ecosystem approaches to conservation. Dysfunctional organisational processes in wildlife agencies were mentioned. A lack of human and financial resources, a trend towards outsourcing work, complicated internal decision-making systems, and priorities driven by short-term election cycles confound consistent and flexible long-term and/or ecological planning. Finally, participants noted that wildlife agencies are not always able to articulate what it is they need from zoos and can enlist zoos too late in the process of recovering native species.

The workshop formulated a draft strategy to address these issues and formulated four sets of objectives. The first two objectives are directed at the zoo community. Participants identified the need to promote better understandings of the recovery process and the potential opportunities this would provide for cost-effective work in zoos and for establishing formal links with wildlife agencies. Objectives directed at zoo/wildlife agency relationships include improving communication among the agencies and increasing wildlife agencies' understanding of zoos' contributions to conservation. Objectives aimed at the species recovery process include gaining a better understanding among zoo professionals of the national and state-level recovery process, and creating a clear and specific process for when zoos begin and terminate their involvement in a recovery project. The ARAZPA will be responsible for the remaining objectives. These include developing a regional planning approach for zoos holding Category One species, designing and implementing training initiatives for zoo staff involved in managing these species, exploring funding opportunities for species recovery projects, and identifying other means for improving zoos' current and future involvement in inter-agency conservation work.

The future success of the ARAZPA workshop will depend in part on the Association's ability to complete—in a timely fashion—production of a formal strategy and other planning mechanisms. Another, more significant challenge will be whether ARAZPA staff and other members of the zoo community are able to continually raise awareness amongst their colleagues about not just the specific issues raised in the workshop, but also the principles on which the workshop was based.

There are other formal institutional structures that can contribute to stronger zoo/wildlife agency relations. The Alice Springs Desert Park in Alice Springs and the Territory Wildlife Park in

Darwin are part of the Northern Territory's Parks and Wildlife Commission. These arrangements can reduce the typical zoo/wildlife agency, *ex-situ/in-situ* conservation divide. At the Desert Park, there tends to be a greater clarity around the different roles that the zoo and other parts of the Commission have in relation to restoring endangered species. That is, there is a strong sense that the Desert Park provides a captive breeding service *for the Commission* and that the Park's role in environmental education is perhaps its more significant strength. Shared understandings and good working and personal relationships between Desert Park and Scientific Services staff enables close and effective co-operation between captive animal managers and field workers, including joint planning for workshops on arid zone species recovery programs held at the Desert Park. In turn, in 1999, senior Desert Park staff were able to organise the workshop dates to dovetail with the ARAZPA annual conference which was hosted by the Desert Park. These arrangements provided further opportunities for zoo staff and wildlife biologists to meet and develop relationships and to explore further opportunities for co-operative conservation work.

Specific recovery programs demonstrate how, why and when zoos and wildlife agencies work well together. Due to extensive clearing for agriculture and urban development the Helmeted Honeyeater, a medium-sized honeyeater, has been restricted to remnant habitat within the Yarra Valley region of Victoria.[34] The Recovery Team for the Helmeted Honeyeater Program has had representatives from seven different organisations, including two universities, several branches of the wildlife agency in Victoria, two NGOs and Healesville Sanctuary. Since the inception of the Helmeted Honeyeater Recovery Program in 1989, specific attention has been devoted not just to increasing the understanding of the species, biology and ecology, but also to the organisation and management of the program itself.[35] A conscious effort has been made to minimise the hierarchy of the Team so that decision making has been representative of the whole group and creative thinking has not been stifled. The Recovery Team was also given considerable authority and freedom to decide on priorities and directions for the project, without having to spend inordinate amounts of time checking back with the parent organisations represented on the project. Team members were originally assigned clear responsibilities and have been able to operate relatively free from the hindrances characteristic of large bureaucracies. For example, rather than be directed by its own adminstration, Healesville Sanctuary staff have managed the captive breeding component of the Program under the direction of the Recovery Team. In addition to these organisational arrangements, interpersonal dynamics have helped the Program's operation. A strong commitment to the common goal of recovering the Helmeted Honeyeater, and historically good personal and working relationships among Recovery Team members have minimised internal group conflicts and increased the Program's overall effectiveness.

The Striped Legless Lizard recovery effort is another example of a well-co-ordinated inter-agency program. The Striped Legless Lizard is found in the native grasslands of south Australia and is considered to be vulnerable, nationally. A multi-disciplinary Working Group, including representatives from the Victorian wildlife agency, the University of Melbourne, the Victorian National Parks Association and the Melbourne Zoo, has overseen the recovery effort since it began in the early 1990s. The program goals have been well defined and responsibilities of the different participants have been spelled out clearly. The Melbourne Zoo has provided invaluable assistance to the program, including captive breeding and public education expertise, and financial and administrative resources. The informal dynamics of the program are no less significant to smooth operations than the formal ones. Long standing personal relations among some of the participants have played a major role in the establishment of shared goals and good communication.

Conclusion

There are many theoretical and practical problems with the notion of a modern Ark. *Ex-situ* conservation—the primary methodology of the Ark—is limited in terms of how well it can contribute to conserving biodiversity due to its high costs, limited scope and an emphasis on crisis intervention. A shadow is cast over the Ark by a lack of space for and a low representation of threatened species, and poor breeding and reintroduction success rates. Finally, the Ark's overall conservation mission can be weakened by particular social and political dynamics within and among the various members of the fleet that go unattended.

Members of the zoo community have certainly made a concerted effort to minimise these problems and capitalise on the Ark's existing potential through the use of sophisticated collection planning and species management programs. They have also guided their 'vessels' to rendezvous regularly with the rest of the Ark fleet. Meetings such as the ARAZPA workshop can facilitate better working relationships between zoos and their respective wildlife authorities.

Comparing the biblical Ark with its modern counterpart, Noah's mission was much simpler. His journey was relatively short when the floodwaters receded and habitats were intact to which his passengers could return. The task of Noah's Ark through the modern zoo is more daunting, given the size and scope of the challenges confronting zoo professionals. At a time of unprecedented rates of species decline and worsening environmental degradation, the imperative to use wisely what limited time and resources are available to zoos grows ever stronger. Like Noah, zoo professionals have to act quickly. But they must also be able to think laterally in order to maximise their contributions to conserving biodiversity. One way they are attempting to increase their relevance for such a task is by applying their research programs to conservation problems.

3

Research for Sailing or Docking?

Once some practice or activity is generally perceived as a routine and continuing part of society . . . people plan on its continued existence, and having laid their plans they stand to lose if that continued existence is not forthcoming . . . Everyone who makes the effort and adjustment necessary to become a scientist bets upon the continuation of science and acquires a vested interest in it.

Barry Barnes[1]

Zoos have a moral obligation to participate actively in the conservation of species . . . The general lack of precise knowledge about the species to be managed hampers successful execution of this task.

Kurt Benirschke[2]

One of the most significant contrasts between the ideals represented by the Ark metaphor and the realities for contemporary zoo practice is the extensive and ongoing destruction of habitats that are meant be the ultimate destination for today's animal passengers, once they disembark from the Ark. In effect, the stakes are much higher. Zoo professionals wishing to maximise their institution's contribution to biological conservation are seeking ways to ensure greater integration of their *ex-situ* programs and their *in-situ* efforts, and to develop further their own capacities in *in-situ* conservation and other forms of conservation outreach.

Zoo professionals also often cite 'research' as an additional function that is central to their modern mission. Research in zoos includes a broad range of activities that are undertaken by professionals inside and outside the zoo community. One common thread linking all these efforts is their focus on zoo animals. The terms 'science' and 'research' are often used interchangeably by zoo professionals—research is meant to further scientific knowledge in zoos with *science* providing the epistemological foundations for research efforts.

Over the last fifty years, zoo professionals have placed an increasing emphasis on applying scientific knowledge and inquiries towards helping to maintain and restore endangered species in particular and biodiversity in general. Yet members of the zoo community are challenged by the task of ensuring that their research supports these conservation goals. Such challenges come from a particular history of research development that has been shaped by the way 'science' has been defined in Western society, by the interests and involvement of the academic community, by the resources zoos allocate to their scientific programs, and by distinctive theoretical, methodological and taxonomic specialties. While these factors help to explain zoos' achievements, they also lessen zoos' effectiveness in solving problems such as the loss of biodiversity

What Kind of Science?

Certain scientific ideas underlie any research effort and are directly relevant to how well that program can ameliorate environmental problems. Members of the zoo community claim in their annual reports, scientific journals, publicity materials or in the media, that their research efforts are increasing, that these projects are neutral tools always applied impartially to different problems as they arise and that these research programs are increasingly significant for conservation.[3] Some zoo research programs are more applicable to modern conservation problems than others. And the relevance of those projects is determined by the context of the research effort. Since the mid-1950s, historians, philosophers and cultural analysts of science have demonstrated that scientific knowledge is subject to varied conditions and influences. The personal, social/cultural and political beliefs of scientists and their institutional settings direct the questions and answers produced by what has long been assumed to be 'objective' science.[4]

The research in zoos today is a product of particular paradigms. A paradigm is a world view or a general perspective that is applied to understanding the real world, and it is based on certain theories, questions, methods and procedures that share central values and themes. Paradigms are very powerful, because they mould perceptions and practices in disciplines in nearly unconscious ways. That is, paradigms have a kind of hidden quality. People rarely think to question them in the first place, so they continue to determine 'what we look at, how we look at things, what we see as problems, what problems we consider worth investigating and solving, what methods are preferred for investigation and action', not to mention 'what we choose not to attend to' and 'what we do not see'.[5]

Western science, within which various paradigms are framed, promotes certain principles of human/non-human interaction in science, which are not entirely appropriate for environmental problem-solving. Twentieth-century scientific knowledge is produced by isolating variables and subjecting them to systematic experimentation, generating information that is meant to be reliable and can be verified. Two central tendencies of these methods are objectification and reductionism. Through objectification, non-human nature is seen largely as an object for human use and benefit, something that does not necessarily have inherent worth. The subject being studied (in many cases it is nature) is also considered to be separate from the observer. Reductionism seeks to break 'nature' down into sets of knowable or observable elements and events. Often the effect is to understand variables in isolation, while complex interactions among variables are neglected or ignored. This scientific method impoverishes our understanding of the natural world because its overemphasis on 'lower level' variables assumes that the whole is no greater than the sum of its parts.[6] Moreover, this stance frequently denies that researchers use any subjective judgements in selecting those parts of the whole that they feel are fundamental and deserving of study.

In fact, scientism portrays scientific knowledge as instrumental and value-free. Since it is supposedly objective, there should not be any moral or ethical problems associated with it. Nor is it meant to promote particular values or issues, rather scientific knowledge should only be applied to resolving technical problems in policy-making settings.[7] Presumably, in this way scientific knowledge is superior to other ways of knowing. Scientism grew from positivism, which originally asserted that the growth of science provided impersonal, rational and objective methods that would enable society to rise *above* other and inferior ways of knowing such as metaphysics, religion and philosophy.[8]

Modern society remains so heavily reliant on positivist ideals because Western science has been organised in particular ways, has promoted certain social norms and has been increasingly put to economic uses. During the Scientific Revolution and in the transition to industrial capitalism, science shifted from a marginal practice conducted in private settings to a very public practice supported by government funding and/or royal patronage. This was a time when scientific knowledge was deliberately being promoted and collected, as evidenced by its appearance in universities. By the nineteenth century the links between science and technological and industrial development intensified when new scientific universities, industrial research laboratories and technical colleges were formed. Science had spread around the globe by World War II and was increasingly being applied to developing industrial commodities. Science was now a dominant feature of Western culture and had fostered it own kind of imperialism. Its claim to being applicable in any setting had assumed a superior stance over indigenous ways of producing and disseminating knowledge, virtually supplanting those perspectives. Science was (and remains) the knowledge of experts, positioned as legitimate and professional within certain societies.[9]

Yet the distinctions between scientific knowledge and policy are equivocal. This model of science becomes problematic in modern contexts when people (knowingly or unwittingly) presume that its overriding purpose is to make precise predictions. Many people believe that science-based predictions are prerequisites to major policy decisions intended to ameliorate or solve the problems of society, and that scientists are supposedly different from other people who participate in these decisions because their scientific input is objective and value-free.[10] This perspective typically results in people focusing on uncertainty, calling for *more science* (or more technology) to reduce any ambiguities and to bring 'the truth' to bear on decision-making. Knowledge gained through scientific method is supposed to have a special, privileged claim to truth (or closer accurate estimations of it) through its claim to 'objectivity'. Again, traditional scientific methods denote objects of valid knowledge as those things having measurable properties, which exist distinctly outside and independent of the observer. Not surprisingly then, knowledge which can be measured and quantified is privileged over qualitative knowledge, which cannot be so readily bounded and counted.[11] Alternative ways of understanding the world and solving societal problems, even alternative perspectives that are gained by observations and experiments, are devalued, in many cases ignored or silenced.

These trends have significant implications for modern problem-solving contexts in general and for zoos in particular. Since the late 1960s and 70s, there have been growing concerns among the general public (and scientists) about science's social values and responsibilities, especially regarding capricious applications of science to certain kinds of technologies. More specifically, these issues include concerns about the physical, ethical and economic results of technology; research conducted on human and animal subjects; inequitable allocations of resources to certain kinds of scientific research; accidents in research; applications of science that aggravate racial, sexual or class prejudices; and unconventional ways of understanding.[12] Some argue that these problems exist and that science has failed to uphold its promise to benefit society as a whole, because of the narrow standpoint of positivism that many basic and applied researchers in the natural *and* social sciences bring to their efforts.[13] Hence, this kind of science is deficient for addressing the growing number of environmental problems, because it is practised without due regard for the relevant ethical, social, political and economic factors.[14]

Positivist science or 'science' is typically associated with research in the physical and biological sciences. These disciplines explore events, processes and relationships within and between elements of the biological and physical world. However, positivist paradigms can be found in applications of the social sciences and the humanities, which examine human individual and social behaviour and culture.[15] In the early to mid-1960s, the social sciences and humanities, long accused of being the poor cousins to the biological sciences, turned to models of science involving causality, objectivity and mechanism as a way to legitimise their inquiries.

This discussion of Western science's philosophical and sociological foundations should not be confused with denying the benefits (both theoretical and practical) that scientific inquiry has brought to society, such as advances in health care, transportation, communication and understanding environmental processes. Yet acknowledging that science is practised within particular social, ethical and cultural contexts begs that we ask the following questions.[16] Whose interests are served by remaining within a model of Western science? How does adherence to particular scientific paradigms for solving environmental and other societal problems shape the knowledge that is produced? To what extent do traditional scientific inquiries truly aid problem-solving, as opposed to merely adding more complexity to difficult and value-laden situations? How are science and policy combined and with what effects?

Developing Science on the Ark

Zoos' contemporary research efforts are largely characterised by a positivist model of biological science, which has a limited capacity to address complex environmental problems. The philosophical and social influences that have created these conditions can be understood by returning to the development of zoos' roles. The zoo's claim to be a scientific institution came early, and a critical catalyst for this imperative was the zoo's transformation from private menagerie to public institution during the early Scientific Revolution and, eventually, with the Industrial Revolution. The assertion continues, although what counts as legitimate 'science' in zoos has changed as different scientific paradigms come to define zoo research.

The founders of the Paris Zoo, and the London Zoo that quickly followed suit, were adamant that the pursuit of (biological) scientific knowledge was (and should always be) zoos' primary objective. They were certainly less than enthusiastic about zoos providing circus-like entertainment for less-learned members of the visiting public. This disdain for entertainment and preference for enlightenment remains at the forefront of contemporary debates in zoo policy.

The emerging science of zoology had found a home in zoos by the mid-1800s. Most anatomical studies of animals had been unsystematic until the end of the seventeenth century. These inquiries were motivated more by a curiosity about what might be discovered, rather than by tests of a general theory about animal physiology or relationships. Following the publication of Linnaeus' influential classification of plants and animals, a new tradition of research and writing about animal physiology and comparative anatomy was established, lending an air of legitimacy to zoos.[17] Then a somewhat competitive climate characterised the push by zoologists to identify and collect as many different and unique species as possible.

The importance of science's formal organisation in society was (and remains) critical to research development in zoos and should not be underestimated. Influential researchers with links to the early zoos were able to focus their research on taxonomy in zoology and biology using a zoo's living and dead animals. Buffon, a Frenchman who was interested in relationships between living forms was a keeper in what eventually became the Paris Zoo. Cuvier, an anatomist who asserted that all structures in animals are correlated, was a professor at the Muséum d'Histoire naturelle and used the Paris Zoo as a laboratory. The success of the London Zoo can be attributed in part to several societies that were being established at the end of the eighteenth and early nineteenth centuries to cater for specialised areas of concern. The Zoological Society, a scientific body, had a brief to advance zoology, as well as establish the London Zoo.

Overall in the mid-1800s, the number of zoos was increasing, many of the more prominent ones (for example, New York Zoo, Philadelphia Zoo) being established by influential citizens who wanted to promote scientific imperatives. Despite their intent, not all these people were equally successful. The National Zoo, established in Washington DC at the end of the nineteenth century, had science in its charter. Yet it did not have its own veterinarian until the 1940s, and no full research and veterinary facilities until the 1970s.[18] London and Paris Zoos' ability to achieve their early, pre-eminent scientific reputations is also attributable to several other factors. These zoos were located in two formidable imperial metropolises. In addition, as capital cities of their countries, London and Paris were home to other national scientific institutions, such as the Royal and Linnaen Societies and several university colleges in London. The zoos also had close links to mainstream scientific institutions, such as the Paris Zoo's relationship with the Museum and London Zoo's administration by the Zoological Society. For some time, a large portion of the London Zoological Society's membership was made up of leading zoologists.

Between the mid-1800s and the early part of the 1900s, research in zoos did not make significant advances. Many academic scientists had lost interest in zoos and aquaria as laboratory settings for research, because most anatomical inquiries had been completed, and the life sciences had not advanced sufficiently to contribute to research on captive animal management.[19] There were other reasons for this atrophy of scientific endeavors in zoos, such as the depressed economic and social contexts that prevailed around the period of the two world wars. The early twentieth century was a time when zoos' scholarly achievements were generally stultified, the focus resting on commercial concerns.

During the 1960s zoos were in another growth phase. Zoo research was still limited to biological studies. However, it was diversifying and eventually led to current studies in environmental enrichment and behavioral biology, to the use of molecular biology in the study of systematics and genetic variation and to the emerging field of conservation biology.[20] Leaders of the zoo community were pushing for greater scientific advances. For example, the

Director of the National Zoological Park in Washington DC, Theodore Reed, was a veterinarian who was intent on revitalising what was supposed to be zoos' original purpose—scientific research. He and the Secretary of the Smithsonian Institution, with which the Zoo was affiliated, were able to agitate for greater support of zoo research.

This phase in the development of zoo research was significant for animal behavioural studies. Ethology, the study of wild animal behaviour, had been growing in popularity during the early part of the twentieth century. Concerns for the welfare of captive animals had been voiced as far back as the late 1800s and early 1900s. It was not until after World War II that Ethology had developed sufficiently and began to be systematically applied to zoo practices in order to improve conditions for zoos' animals.[21] Influential figures with strong bases of institutional support were able to advance their cause.

Heini Hediger, of the Zurich Zoo, published *Wild Animals in Captivity* in 1950, demonstrating his and others' awareness that zoos needed to make better provisions for captive animals' physiological and behavioral needs. Training and play were being applied as a sort of occupational therapy in an effort to achieve the kinds of behaviour that field biologists observed in wild animals.[22] Hediger, an early advocate for zoo-based research, was also voicing his concern about the low priority research received in zoos and promoting the need for strong links with universities as a way to facilitate scientific advances.[23]

Yet Hediger and other prominent individuals were not satisfied with waiting for universities to take the lead. Desmond Morris who had studied with the accomplished ethologist, Niko Tinbergen from Oxford, started the first formal zoo behavioural research unit at the London Zoo in the early 1960s.[24] The New York Zoological Society quickly followed suit in 1965, opening a research institute in collaboration with Rockefeller University. Interestingly, neither unit survived after their respective founders left the Zoos. The National Zoological Park set up its research unit in 1965 as well, hiring the mammalian ethologist John Eisenberg. This represented a departure from the research trends in prominent American zoos (such as Philadelphia and San Diego) that highlighted physiology, nutrition and genetics.

The focus of contemporary zoo research came about during the late 1970s and 1980s. Again, the latter half of this century was the period when zoos' conservation ethos appeared in full. The term 'biodiversity' was coined in the mid-1980s, and is now widely used to refer to the richness and variety of life on earth and to an area of scientific research that describes, measures and seeks to preserve diversity and its sources.[25] Leading up to this time, the scientific community, parts of which had links to the zoo community, was producing convincing estimates of the rates of biodiversity's decline through tropical deforestation and the attendant loss of habitat where such a high number of species are concentrated.[26] This problem of extinction was being noted in other ecosystems as well. Conservation biology was partly a response to what was seen as a *biological* crisis and was also informed by rapid advances in the biological sciences.[27]

This emerging discipline also had an applied focus, seeking to bridge a gap that an increasing number of scientists (and others) perceived between pure science and 'practical' management. The discipline grew from an array of concerns and activities. Zoos were trying to establish and maintain small populations of endangered animals; botanic gardens and agricultural departments were seeking to preserve shrinking genetic resources of plants; forestry and fisheries services endeavoured to maintain a maximum yield from the resources they managed; conservation agencies were trying to start and protect nature sanctuaries; and biologists were attempting to explain declines in the species they studied.[28] High profile members of this research community, such as David Western of Wildlife Conservation International, were promoting the fact that conservation biology would provide a range of criteria for recognising biological boundary limits, identifying species or ecosystems in decline, and distinguishing critical ecological processes and keystone species. They were also making a push to move away from the narrow notion of wildlife management to the broader concept of conserving biodiversity.[29]

The disagreement about what conservation biology should focus on remained largely centered on biological issues and illustrates how, even within a single discipline, there can be competing scientific paradigms. By the mid-1990s, Graham Caughley, another prominent scientist, had described two paradigms that he believed were (and are) driving most of the research activities in the relatively new discipline and that needed to be more fully integrated.[30] The small population paradigm typified early work in conservation biology when it emerged as a scientific discipline in the 1980s. This collection of ideas deals with the effect of smallness on the ability of a population to survive (that is, the risk of extinction inherent in low numbers). Intensive population management was the method chosen for accomplishing a conservation mission (that is, species conservation through long-term retention of genetic diversity that provides a safety net for wild populations if reintroduction is needed in the future).[31] It has a strong focus on population genetics and population dynamics problems faced by a population at risk of extinction because its numbers are small and those numbers are capped. *Ex-situ* conservation had found a logical home in the zoo community, whose members were already hard at work developing this expertise.

Caughley noted that, in contrast to the small population paradigm which deals primarily with the size of a particular population, the declining population paradigm focuses on the *(biological) causes* of that smallness. That is, it looks for those external processes that drive a population towards extinction and for what can be done to halt that decline. Caughley suggested that the small population paradigm's limitation was that its users believe that identifying the problem's causes is beyond the paradigm's brief in the first place. Other scientific standpoints competing for attention have produced similar and related criticisms of, not just the small population paradigm, but the institutions in which it is embedded, namely zoos. As we saw in Chapter 2, those who endorse holistic approaches to conservation critique *ex-situ* conservation on the basis of problems that occur with establishing self-sufficient captive populations, low reintroduction success rates, domestication and disease outbreaks. Additionally, *ex-situ* programs (and their associated research activities) are constrained by high costs, the need for intensive management and high levels of inter-agency cooperation and zoos' finite organisational, financial and spatial resources. Finally, some believe that the species approach distracts research and public attention away from the importance of habitats, ecosystems and ecological processes.

Despite these critiques, conservation biology and other scientific disciplines practised in zoos and in other environmental institutions remain largely embedded in a positivist tradition. Again, this paradigm tends to objectify nature, positions people as separate from nature, assumes a value-free stance, attempts to predict with accuracy, and privileges quantitative knowledge and methods over other ways of knowing and problem-solving. Research in the physical and biological sciences is favoured, while support for qualitative social inquiries languishes. Open discussions about the social and political contexts inherent in dominant research paradigms seem to be neglected. In recommending future directions for conservation biology, Andrew Burbidge, a scientist at the Department of Conservation and Land Management in Western Australia stated:

> To be successful, conservation biology must produce the information needed by conservation policy makers, bureaucrats and managers, and be applied. The challenge for conservation biologists is to meet needs in systematics, ecological survey and monitoring, the management of individual threatened taxa (including ex situ conservation where appropriate), exotic species control, conservation genetics, ecosystem management, and conservation outside reserves.[32]

This standpoint assumes that with ever increasing amounts of biological knowledge we will be uncategorically able to decide what parts of nature to protect, how to avoid extinctions and how we can restore ecological damage. It remains largely devoted to reducing uncertainty by calling for precise, quantitative models of research that will provide numerous predictions for solving problems. There have certainly been significant advances in environmental circles in understanding biophysical processes and in applying that knowledge to technical and management options. However, there is movement afoot to show that this information has not been sufficiently disseminated, adopted or implemented, that environmental degradation is therefore continuing and that the influence of social, political and economic contexts on environmental decision-making must be incorporated into conventional research and management agendas.[33]

Institutional and Social Influences on Zoo Research

Certain institutional and social structures help to maintain zoos' traditional research priorities and determine the size and duration of programs. The amount of human and financial resources allocated to research will influence how extensive the work will be and who will undertake the bulk of those activities. Obviously, zoos have to make choices about what to emphasise or specialise in given their finite resources. In addition, research directions tend to follow the particular competencies and interests of the people involved.[34] Researchers in zoos are typically trained in the biological sciences: they include zoo staff, professionals and students from universities or some combination of the two. Zoos' relationships with certain parts of the university community are not only critical to their ability to conduct research, but these connections determine the kind of work that takes place.

Between 1997 and 1998 two major studies detected some discernible trends in research and publication activities, primarily in the North American zoo community. Chris Wemmer, of the National Zoological Park and her associates analysed articles published in a prominent zoo industry journal, *Zoo Biology*, published in the United States by the American Zoo Association.[35] They found that 43 per cent of senior authors were from the zoo community and another 26 per cent from universities. Nearly a third of the articles resulted from some form of collaboration between zoo and university personnel or students; another third were just from the university community. Zoo professionals wrote the remaining third. Not surprisingly, affiliations between zoos and universities or museums appear to encourage formal research in zoos. Tara Stoinski and her colleagues from Zoo Atlanta and the Georgia Institute of Technology found that 83 per cent of zoos affiliated with universities and 71 per cent of institutions associated with museums or other research organisations reported publishing their research findings in refereed journals.[36] Interestingly, those numbers dropped considerably where no such affiliations were reported.

It would also appear that certain institutions dominate research and publication activities. Wemmer and her associates found that the senior authors of almost half of the articles published in *Zoo Biology* were staff from zoos belonging to the American Zoo Association (AZA). This is not entirely surprising given that the journal is published by the AZA. In any case, these institutions represented only 14 per cent of the 176 institutions belonging to the AZA. Over half of the articles written by senior authors from AZA zoos came from four of the more prominent members of the American zoo community. These included the Wildlife Conservation Society in New York, Zoo Atlanta in Georgia, San Diego Zoo and Wild Animal Park in California, and the National Zoological Park in the nation's capital. Similarly, Stoinski and her colleagues found that respondents to their survey also listed well-established organisations when identifying zoos with the best scientific reputations. The Wildlife Conservation Society, San Diego Zoo and Wild Animal Park, and the National Zoo reappeared at the top of this list. Additionally, other prominent zoos and aquaria were listed, such as Brookfield Zoo in Chicago, St. Louis Zoo in Missouri, Cincinnati Zoo in Illinois, Monterey Bay Aquarium in California, the National Aquarium in Baltimore and others. Wemmer and her colleagues suggest that the dominance of four zoos as contributors to over half the scientific articles in *Zoo Biology* could be explained by the possibility that most members of the zoo community are not as seriously engaged in research as some might claim.[37]

Those zoos that do excel in research appear to have substantial bases of institutional support. The London Zoo has undertaken research in genetics, endocrinology, and biochemistry through its direct association with the Institute of Zoology. The Philadelphia Zoo established the Penrose Laboratory, now a major research facility producing nutritional, microbiological and other work. The New York Zoological Society operates the Wildlife Conservation Society, which focuses exclusively on field studies of the ecology and general biology of endangered animals. San Diego Zoo operates its own research department. The affiliation between the

National Zoo and the Smithsonian Institution has led to extensive biological research on captive and wild animals and the publication of that research through the Institution's in-house publishing company.

Why are zoos' links to the university community so important? As we saw, these relationships have provided the foundations of research activities in zoos, scientists realising their research interests in part through or with zoos. Today, limited human and financial resources are allocated to zoo research programs. Tara Stoinski's survey found that for those zoos reporting that they did not conduct any research activities, financial constraints (80 per cent), time constraints (70 per cent) and a lack of qualified personnel (40 per cent) were most commonly listed. In that same study, 84 per cent of zoos undertaking research reported funding their research through their operating budgets, 94 per cent reported that research made up 0–5 per cent of that budget and 74 per cent were dissatisfied with this level of funding. Most zoos simply have not devoted sufficient resources to maintain comprehensive research programs and therefore require the support of other institutions. Yet the type of resources available to and administrative environments in zoos may make it difficult to forge links with university researchers. Few zoos, save for the most prominent and well-funded organisations, have earmarked budgets, facilities or formally-trained staff for the kind of laboratory-based research that is required of particular scientific disciplines.[38] The work culture of a university environment and standards for measuring success are likely to be different from those of a zoo and might deter scientists from working in or with zoos.

Not surprisingly, most research in Australian zoos also tends to come from the larger, better-funded zoos. Since the late 1990s, three of these—Perth Zoo, the Zoological Parks and Gardens Board of Victoria and the Zoological Parks Board of NSW—have designated separate administrative units that are responsible for undertaking and co-ordinating research activities. Again, certain social patterns and administrative arrangements partly determine the amount and kind of research activities. Some of the major Australian zoos have strong links with biological researchers in the academic community and/or their respective wildlife authorities. Convivial personal and professional relationships were demonstrated to be critical to zoos' involvement in endangered species restorations programs as we saw in Chapter 2. These same dynamics are just as critical to maintaining viable and productive zoo research programs.

Perth Zoo has traditionally had strong links with the academic community. Twenty-seven of the forty-seven current and completed research projects listed in the Zoo's research review for 1996–98 involved some level of collaboration with university-based researchers. Historic associations with universities in the Perth region have enabled undergraduate and post-graduate students to undertake an array of biologically-based studies on the animals held by the zoo (for example, Murdoch University which has a prominent veterinary school, the University of Western Australia, etc.). Much of the research work at Perth Zoo is linked to the Co-operative Research Centre for the Conservation and Management of Marsupials. CRCs are collaborative research programs among professionals from industry, research organisations, educational institutions and government agencies. CRCs are partly funded by the federal government; the sixty-seven programs currently operating receive an average of $2.2 million per year, with the expectation that participants will match that amount with their own funding. The CRC operates a Native Species Breeding Program at Perth Zoo and participants include Macquarie University, Western Australia's Department of Conservation and Land Management, Western Australian Museum, University of Western Australia, Murdoch University and the University of New England. The past and current Directors of the Perth Zoo component of the Marsupial CRC are university professionals with extensive (and specialised) experience in biological research.

Perth Zoo's high profile in conservation research hinges on the links it has been able to establish with the Marsupial CRC. Funding issues are critical to the longevity of the zoo-based program. The CRC must reapply to the Federal government funding on five-year cycles, the Zoo paying a contribution of total costs (that is, salaries for the Zoo Research Director and two keepers in the Native

Species Breeding Program). There has been considerable speculation among environmental policy and research circles that the present Liberal Federal government's interest in environmental research has waned, and that those CRCs that undertake this kind of work are in a more precarious position than other CRCs whose work is more commercially-oriented. It is not clear what would happen to the Zoo's research activities if the larger CRC were disbanded due to a lack of funding. Presumably, other research activities undertaken at the Zoo that are not tied to CRC funding would survive.

In NSW, the Zoological Parks Board houses a Conservation Research Centre at Taronga Zoo. As of September 1998, the Centre listed forty-eight research projects that were being conducted at Taronga and Western Plains Zoos. Nineteen of those projects involved researchers from universities. The Animal Gene Storage Resource Centre is also based in this unit. It involves research collaborators from the Institute of Reproduction and Development at Monash University, the Zoology Department at the University of Melbourne, the Wildlife Reproduction and Management Unit in the Faculty of Veterinary Science at the University of Queensland and the Department of Animal Science (Veterinary Science) at the University of Sydney. The Animal Gene Storage Resource Centre of Australia was founded largely through the efforts of Jack Giles, Director of Scientific Policy & Research for the Zoological Parks Board of NSW, and Alan Trounson, Director of Monash University's Institute of Reproduction & Development. The two share an interest in artificial reproduction and worked with a group of physiologists, geneticists and biochemists from several universities and the Director of the Zoological Parks Board of NSW to formulate the original project proposal.

Again, funding for research activities is an important consideration. The Zoological Parks Boards openly designates research (and conservation) activities as 'discretionary', therefore needing to be 'carefully planned and largely funded from grants, sponsorship and other sources of dedicated income, distinct from core funding'.[39] The ability of the Board to sustain its two main research departments, the Conservation Research Centre and the Australian Conservation Training Initiative, has been predicated upon its capacity to draw in outside funding, as 'our objective is to achieve full self-funding'.[40] There are implications in these kinds of arrangements, as we will see. Generally speaking, the greater the dependency of research and conservation activities on external monies, the more susceptible they are to being determined or influenced by the political and scientific interests of the funding bodies.

The Adelaide Zoo and the Monarto Zoological Park are overseen by the Royal Zoological Society of South Australia and have long-standing associations with several universities in the Adelaide area. The current chairman of the Zoo Board, Mike Tyler, is also an Associate Professor in the University of Adelaide's Zoology Department. Other recent Board members have belonged to the academic community. Both Professor Tyler and the Society's CEO, Ed McAlister have considerable standing in the local and national university, conservation and governmental communities. They are well positioned to facilitate research and conservation-oriented research activities in both the Zoos. In addition, senior curatorial and veterinary staff, as well as staff from animal management units have strong relationships with staff in the current State government's wildlife agency and have undertaken joint conservation research work. Numerous students from zoology and psychology departments at the University of Adelaide and Flinders University have undertaken studies of the Zoos' animals. In 1998, the Society approved eight research projects to be undertaken by professionals and students from the university community. Presumably these links with local universities and government wildlife authorities make up the gap between the Zoos' interest in research and the lack of funds, especially since the Zoos do not have specially designated conservation research units. The Society reported that it spent just over $5000 for conservation research activities in 1997.

The Zoological Parks and Gardens Board of Victoria also has strong links with the academic community. The Zoo Board's scientific advisory committee has included representatives from the Faculty of Science and Technology at Deakin University, the Faculty of Veterinary Science at the University of Melbourne, the Department of Genetic and Human Variation at La Trobe University, the Department of Obstetrics and Gynecology at Monash University and the Department of Prenatal Medicine at the Royal Women's Hospital. Some of these people have also sat on the Board's animal experimentation committee and/or serve as scientific associates for the Zoo Board. Two other appointments show more clearly the facilitative role that close links between zoos and universities play. The previous head of Melbourne University's Zoology Department, Angus Martin, now acts as a chief consultant for the Zoo Board and maintains an office in the Board's Conservation and Research Unit at Melbourne Zoo. The Director of the Conservation Unit, Peter Temple-Smith, is also an accomplished scientist, previously serving as a Professor at Monash University in Victoria. Through their strong links with the university community and with Melbourne University in particular, the Conservation Unit has been able to establish laboratory facilities for the Zoological Parks and Gardens Board in the Zoology Department. The Zoological Parks and Gardens Board's annual report for 1998 lists thirty-nine separate research projects. Zoo staff conduct 36 per cent of those projects, university professionals and students undertake 20 per cent, collaboration between zoo staff and university professionals and students government agencies and NGOs accounts for another 20 per cent and the remainder of projects.

Since the late 1990s, the ZBV has provided fairly extensive support for its research unit. The Zoo Board covers salaries and administrative overheads. The Unit's Director negotiates a minimum amount of external funding that his staff must obtain. However, given that there are always more projects than there is money for, the Unit pursues far more funding than the agreed upon minimum. Here again, strong relationships with the academic community extend the Unit's capacity to secure funding and take on new projects. For example, in 1999 the Unit was part of a successful application which the University of Melbourne's Zoology Department submitted to the Australian Research Council for research on marsupial reproduction. As a result, the Conservation Unit will be able to draw on the University's staff and facilities for their research.

The Ark's Research Categories

Different kinds of research are undertaken in zoos. Some research is *about zoo animals*, some involves *using zoo animals*, while other research goes *beyond the zoo* either as *in-situ* research or in drawing on related fields such as social science. Zoo staff, personnel from universities and from government agencies conduct research, either separately or in collaboration. Research *about zoo animals* and *beyond the zoo* tends to be applied: it has an emphasis on solving specific problems. Research *about zoo animals* generally focuses on individuals and species in a zoo for the purposes of maintaining the animal collection in perpetuity. One example would be veterinarians studying ways to prevent and/or treat disease among a zoo's population of captive animals. Another example would be behavioural studies that aim to provide environmental enrichment for captive animals. Members of the academic community, zoo staff (for example, a zoo veterinarian) or a combination of the two, typically perform this kind of research. Research *beyond the zoo* tends to consist of activities that are associated with conserving wild populations of animals, some of which may be held in zoo collections for conservation purposes. This category can also include social science research, such as studies of the role of zoos or of people's attitudes to animals and their learning experiences in zoos. Zoo staff, university personnel and students, or a combination of the two may also undertake these inquiries.

The third type of research, *using zoo animals*, is conducted by members of the academic community and would be the least relevant to zoo activities or conservation *per se*. It need not have any applications to animal maintenance or be of interest to the zoo community. It is more oriented to pure or basic research; it tends to be driven by the desire to test theories and to increase biological knowledge.[41] This research relies on using biological materials taken from live or dead zoo animals for physiological studies of one kind or another. For example, zoos are typically approached by university scientists to use tissues from necropsies, blood from animals and other materials for research on the mechanisms of ageing, taxonomy and biochemical studies.[42]

These divisions are not always as neat and clear as they might appear. Sometimes a research project will be mostly focused on ensuring the health of the captive population, but findings may have some unexpected applications for managing wild populations of a species. Conversely, knowledge of wild populations can be used to improve the quality of life for captive species. Research that mostly uses zoo animals as experimental specimens may have some applications to conservation, such as assisted reproduction research. The successful birth of the gorilla 'Mzuri' at Melbourne Zoo was achieved by applying IVF technology developed for humans. Many members of the zoo community today are increasingly concerned with research that has applications for conservation issues 'outside' the zoo.

In addition to the three broad categories—*about zoo animals, beyond the zoo, using zoo animals*—zoo research activities are organised around particular themes and scientific disciplines. Remembering that research specialties are determined not only by available resources and the expertise and interests of individuals, but also by the needs of specific groups of animals held in the zoo, zoo research is (and has been) most thoroughly dominated by positivist models in the biological sciences. In addition, virtually all research has focused on individual species, although it has been noted that zoos concentrate their limited resources on fewer types of studies and smaller numbers of species. It would appear that basic research on common species has clearly been less appealing to zoos than applied research on endangered species.[43]

Wemmer's analysis of articles published in *Zoo Biology* certainly provides some further evidence of these biases. She found that reproduction and behaviour were the predominant themes in published research findings with relatively few on genetics and demography. These latter two disciplines are important to small population biology research, which informs planning for *ex-situ* and *in-situ* conservation. As Table 4 shows, Stoinski also found that behavioural and reproductive studies are prominent in North American zoo research.

Table 4 Research themes in North American zoos

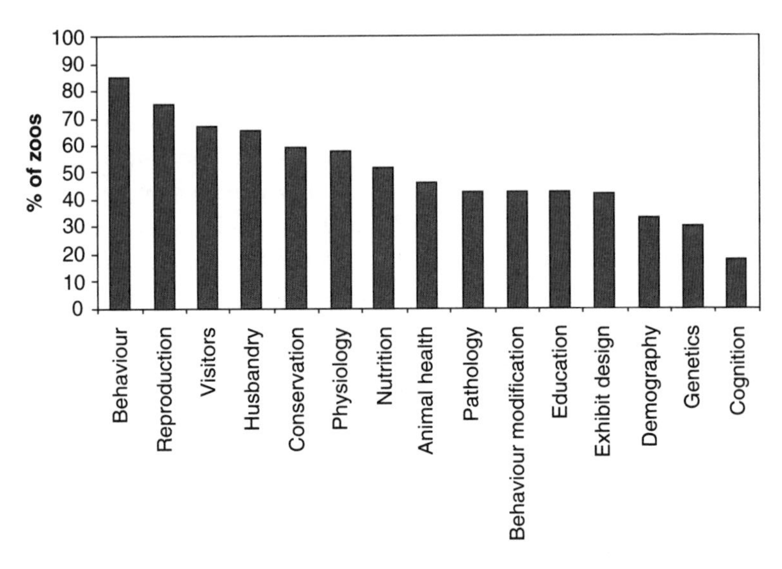

SOURCE: T. S. Stoinski, K. E. Lukas and T. L. Maple, 'A Survey of Research in North American Zoos and Aquariums', *Zoo Biology*, vol. 17, 1998, p. 171.

Traditional zoo animal collections have highlighted charismatic mammalian fauna. Wemmer's and Stoinski's studies found similar biases in the research foci at North American zoos. Wemmer and her colleagues showed that a resounding majority of research *articles* focused on mammals. Stoinski and her collaborators' analyses of zoo research *projects* found a comparable bias towards mammals. Here the zoos most commonly reported that their research staff studied carnivores, hoofstock and primates, such as the great apes. Many of them also concentrated their research activities around small mammals and marine mammals. Stoinski's study also found that over half the zoos responding to the survey reported that their research focused on 'people'. Visitor evaluation studies tend to come mostly from the United States (see Chapter 4), and people-oriented research in Australian zoos tends to be marketing-based. Given that Stoinski did not report on specific descriptions of zoo research programs, it seems likely to assume that a fair percentage of the 'human oriented' research zoos were reporting on what could have been market-oriented analyses which try to determine zoo visitor preferences and habits.

The kinds of research practised in the major Australian zoos are similar to research trends in North American zoos (see Appendix 1). My analysis focused on several different kinds of publications: conference proceedings, newsletters/magazines, book chapters, academic theses and journal articles. The major Australian zoos also list their major formal research projects in their annual reports, providing another indicator of research trends. As with their North American counterparts, the dominant research theme is the biological sciences and behavioural and reproductive studies of a limited number of species and of mammals in particular. Not surprisingly, marsupials and monotremes, the platypus especially, are common subjects for behavioural, reproductive and genetics research.

Zoo Research: Conserving Captive and Wild Populations?

Modern zoo policy imperatives urge members of the zoo community to undertake and support research in their organisations. The World Zoo Conservation Strategy states that 'much of the information acquired through zoo research is of great relevance to conservation generally and to the conservation of species and habitats in particular'.[44] As stated earlier, the first two broad categories of zoo research, *about zoo animals* and *beyond the zoo*, are the most relevant to the zoo conservation role. The projects that qualify as research *beyond the zoo—ex-situ* programs undertaken in support of or directly linked to *in-situ* conservation—are valued most highly by zoo professionals seeking to expand zoos' conservation potential, as indicated by the World Zoo Conservation Strategy:

> Although zoo research will not be directly applicable to conservation of wild populations in all cases, it can often stimulate and direct research undertaken in the wild. Ex situ and in situ research are thus intertwined . . . Many zoos are also involved with conservation oriented research in the field, within their own local region, or elsewhere in the world. They therefore contribute directly to the increase

> in knowledge concerning conservation of species and habitats. It is an important objective of the World Zoo Conservation Strategy to stimulate further this type of involvement of zoos, as it forms a valuable link between ex situ and in situ conservation efforts, as well as between single species conservation and ecosystem conservation.[45]

Much of the formal research conducted by the major Australian zoos falls under the categories of research *about zoo animals* and research *beyond the zoo*, and has indirect and direct relevance and applications for conservation objectives. These projects are also exclusively biological studies that primarily emphasise single species and rely on highly technical, quantitative information.

Research Beyond the Zoo

The model held up as the ultimate in zoo conservation achievements is the work of organisations such as Wildlife Conservation International, a division of the New York Zoological Society. Wildlife Conservation International has gained considerable prominence, because it focuses exclusively on *in-situ* conservation work, combines conservation biology and ecological studies and includes social and economic issues by linking its research efforts with wildlife and conservation programs of the local communities. Wildlife Conservation International study sites are located primarily in conservation 'hot spots', places in non-Western countries that are home to high levels of biodiversity such as Central and South America, Africa and South-East Asia.

Approximately 11 per cent of the Zoological Parks Board of NSW's research projects during the late 1990s focused on *in-situ* work. One study of Little Penguins has tried to determine the impacts of urban and agricultural run-off flow in the waters surrounding an island which is home to a colony of these penguins. Another zoo-sponsored *in-situ* project examined the population dynamics of feral cats in order to develop more controls of this destructive introduced pest. The Director of the Conservation Research Centre has undertaken some work on controlling feral pigs in several regions of NSW.

The amount of *in-situ* conservation research work that has been done in the last several years by the Zoological Parks and Gardens Board of Victoria is slightly higher, accounting for approximately 20 per cent of its total research projects. These research efforts include monitoring macropod populations in a reserve that is a release site for an endangered species of bandicoot, population surveys of an endangered species of moth and the biology and reproductive behaviour of wild platypus. Similarly, roughly 19 per cent of the Royal Zoological Society of SA's research projects during the late 1990s have been field-based. For example, zoo staff have monitored population fluctuations of the endangered Yellow-footed Rock Wallaby at a regional release site. Another *in-situ* study involved testing the effectiveness of biological control for a particular introduced plant.

Research about Zoo Animals

A more common form of research is that work which is based on captive populations of animals but has fairly obvious links to (and benefits for) conserving particular species in the wild. In Victoria, approximately 26 per cent of the Zoo Board's research programs in the late 1990s were part of a broader, on-going *in-situ* conservation effort. For example, as a member of the Helmeted Honeyeater Recovery Team, Healesville Sanctuary staff have undertaken research on captive population management and reintroduction strategies, as well as wild population monitoring and management for the species. Work by Sanctuary staff on cross-fostering of Helmeted Honeyeater eggs with analogue species was critical to the successful establishment of a founding captive population, which could then be used to supplement wild populations.[46] Sanctuary staff resolved questions about how to track released birds by developing and trialling methods for attaching radio transmitters to the birds and for releasing those devices. Nutritional studies enabled staff to redress some early problems with the captive population, such as loss of pigment in the bird's plumage, fragile eggs, soft bones and muscular weaknesses. The Melbourne Zoo's Veterinary Department and the State wildlife authorities studied the incidence of a particular disease among Eastern Barred Bandicoots that had been bred at the Zoo and released into a nature reserve. Zoo and government agency staff were able to determine that disease was probably not contributing to the decline of the wild population. In a similar partnership, veterinary staff assisted wildlife agency personnel in their ecological study of Carpet Pythons by surgically implanting radiotelemetry devices in the animals prior to their release back into their wild habitats.

In NSW, roughly 15 per cent of the Zoological Parks Board's research in the late 1990s focused on captive animals that have been and are used for reintroduction programs. For example, zoo staff have conducted experimental releases of species such as the Western Hare Wallaby, the Greater Bilby and the Mallee-fowl. Western Plains Zoo staff have been working with the NSW National Parks and Wildlife Service to establish a suitable habitat for the Mallee-fowl in a nearby reserve and to restock this area to a self-sustaining levels with birds bred at the Zoo.[47] Adelaide Zoo and Monarto Zoological Park in South Australia, in conjunction with independent researchers, are also involved in the National Recovery Program for this species. Monitoring studies of newly hatched chicks released into Monarto aim to improve monitoring of the species in the wild. This type of research comprised an estimated 38 per cent of projects undertaken by Adelaide Zoo and Monarto Zoological Park staff in 1997 and 1998. One of the zoo veterinarians studied the incidence of cataracts in the endangered Stick-nest Rat and was also involved in field studies on this species. Zoo staff have also participated in studies on the reintroduction of certain threatened plants.

In Western Australia, 15 per cent of the Zoo Board's research from 1996–1998 focused on threatened species, such as the Numbat, Chuditch, Dibbler and Shark Bay Mouse. The Perth Zoo is a member of the Numbat Recovery Team. Therefore, zoo staff conduct research on the reproductive biology of captive Numbats to ensure that there are sufficient numbers of animals available for the release component of the recovery effort. Similar work is undertaken for the Chuditch Recovery Plan. Perth Zoo staff work with Western Australia's Department of Conservation and Land Management to monitor the health of wild and captive populations of Chuditch. These activities include collecting blood samples from individuals in the captive colony and individuals caught during field surveys to determine normal values, as well as for DNA analysis.[48]

There are other types of projects that fall into the broader category of research *about zoo animals*. They are similar to the research efforts just described, but it can be said that their links to *in-situ* conservation are more ambiguous. These projects tend to utilise expensive, high technology and fairly invasive methods to achieve their ends. Questions about these projects focus on how practical they are for restoring biodiversity in wild settings and on broader ethical and moral issues associated with the appropriate use of technology in society.

For example, the Zoological Parks Board of NSW's Conservation Research Centre hosts the Animal Gene Resource Centre of Australia. Research in the Centre since the late 1990s has been oriented towards conservation of highly endangered species in several taxonomic groups such as monotremes (Long-beaked Echidna), marsupials (Northern Hairy-nosed Wombat and others), ungulates (Black Rhino), Aves (Black-eared Miner, Mallee Fowl), and carnivores (Snow Leopard, Asiatic Cat, Sumatran Tiger). Many of these species are already part of the Zoological Parks Board of NSW's animal collection. Work is being conducted on the collection of reproductive 'resources' (for example, live semen and eventually unfertilised eggs and embryos), cryogenic storage of sperm, eggs and embryos, and assisted reproductive technology (for example, artificial insemination using fresh and frozen semen and test tube fertilization). The Centre's research brief is meant to (eventually) eliminate the risks of inbreeding and loss of genetic fitness that can occur after long periods in captivity, and to reduce the costs of maintaining large numbers of animals in captivity by maintaining high volumes of reproductive 'resources' (that is, sperm, eggs, embryos).

One question worth considering is the degree to which these advanced technologies further *ex-situ* conservation objectives as opposed to achieving *in-situ* conservation. The Centre justifies its research agenda on the basis that 'some species will become extinct in the wild because threats to their survival cannot be removed, at the present time at least', and that 'captive breeding is the only way such species can survive'.[49] The tangible prospects of reintroduction in these cases are highly ambiguous, the habitat destruction (as opposed to the causes of it) are almost accepted as given and/or outside the project participants' briefs. Just what is meant by 'survival', captivity in perpetuity or *in-situ*, is unclear.

The Black Rhino is a high-profile case where this logic has been applied. Certainly the Centre's work and the genetic studies undertaken on this species (and of various other zoo animals) provides valuable knowledge for maintaining the species' genetic fitness while they are kept in captivity. However, such efforts tend to be presented publicly as being *critical* to the species 'conservation' when the reality is that the wild habitats of all species of rhinoceros are far from being assured and efforts to conserve them are fraught with political, economic and social complexities.[50] One might well ask for clarification on what kind of conservation is being practised by programs such as the Zoological Parks Board of NSW's importation and maintenance of and experimentation with Black Rhinos, and for what purposes?

The reductionist, scientific reasoning underlying many of these high tech *ex-situ* programs is problematic, because it assumes that, as in building a machine, a functioning system can be reproduced (the ecosystem rhinos live in) by assembling an adequate number of component parts (the rhinos). Yet habitats are far more complex than the sum of their total parts, and species such as the rhinos are more than their genomes. They have certain behaviours and social lives, and must interact with their environment to survive. The partial success of reintroduction programs is testimony to the fact that even where sufficient individual animals can be assembled, ensuring that they can be incorporated back into natural communities is quite another matter.[51] Zoos have a responsibility to maintain animals in conditions where as many of their natural behaviours as possible can be reproduced. A more appropriate zoo response to the rhinos' plight might be to devote more resources to maintaining the species in conditions close enough to 'wild' settings that they can breed and rear their young, without the assistance of expensive, high-tech assisted reproduction methods. Moreover, if zoos must maintain animals for long periods of time before they can be reintroduced, then the duty of care is to maintain them as fully functioning animals, not as genomes, especially if you consider that this might be the only life they will ever have.

Not surprisingly then, some conservation biologists warn that while *ex-situ*, '. . . high tech methods . . . give the field glitz and public appeal', mastery and applications of technology should not be misrepresented or misread as conservation success.[52] These sentiments are really a microcosm of broader debates about the social responsibilities of science mentioned earlier and about the appropriate development and use of technology in society. There is considerable recognition that, while technology has significant benefits, it also has a demonstrated capacity to generate important and often unexpected side effects.[53] Nonetheless, like science, technology is also promoted and sustained by social, political and economic systems that are not particularly conducive to candid, participatory assessments of its use. The undesirable side effects of technology catch people off-guard, often because sufficient debate about its potential ramifications have not taken place early enough or have been delayed in the process of developing and applying that technology.

Are there extensive and open debates about the implications and uncertainties associated with the high-tech methods zoos use for *ex-situ* conservation? Certainly zoo professionals have engaged conservationists and philosophers in discussions about the ethics associated with using zoo animals in research and about the effectiveness of *ex-situ* conservation. The more prominent defenders of zoo research tend to assert publicly that their ethical duty of care to captive animals extends to minimising suffering and loss of life. The extent to which they move beyond that issue to question the underlying paradigms of either the science, or the technology informed by that science, seems to get caught up in defensive and restricted logic. For instance, several prominent zoo officials said, 'like it or not, the survival of many species, especially the larger vertebrates, is going to require unprecedented levels of intervention', and a range of intensive management actions 'will become increasingly necessary'.[54] They argued that, consequently, evolving technologies such as embryo transfer, artificial insemination, cryo-preservation, are and will become *more* important tools for conservation.

The need for intensive and technological intervention in conservation is presented as absolute, the only sensible thing to do given the gravity of the situation. The effect of this position is to leave little room for seriously considering *other* kinds of intervention or the appropriateness of science and technology and any unforeseen and/or undesirable consequences that might arise from its use. Yet, technological knowledge does not 'cover all the bases', it does not translate automatically into social wisdom simply because it is developed by specialised expertise.[55] For example, the issue of proprietary and ownership rights of the animal material stored in genome banks warrants further discussion.[56] In addition, vast amounts of human, animal and financial resources are required to work out the kinks of gene storage, cloning and assisted reproduction techniques, while, in the interim, vast numbers of species and habitats are lost every day. Some private corporations are eager to lend their financial support to biotechnology research, but they are also keen to reap the profits from the sale of that technology once it is developed.

These questions represent the kind of knowledge that tends be neglected by the positivist reductionism of technological knowledge. Particular philosophical, social and institutional structures sustain such conventional models of science and technology, insulating some practitioners from these debates. In other words, members of the zoo and academic research community who have devoted considerable amounts of their time, energy and resources to developing and implementing high-tech *ex-situ* conservation projects, may feel threatened by and resist questions about the appropriateness of their favoured methods and technologies.

Old-fashioned Research for Modern Problems?

Overall, the amount of research in zoos has grown over the years and shifted increasingly to applied research. Members of the zoo community, both in Australia and overseas, are making concerted efforts to apply their research efforts to conservation problems.

We have also seen that zoo-based research is biased towards particular scientific disciplines and animal taxa, and that these leanings can traced to the social, technical, and philosophical orientation of traditional scientific knowledge and methods. Exclusive use of traditional scientific knowledge and methods has been less than effective in reversing local, national, and global environmental degradation.

Conservation biology is an important discipline for zoo research, partly because its professional culture is talking about the possible need for a new scientific paradigm. Some conservation biology practitioners have always been more forthcoming about the ideological underpinnings of their discipline than those in older, more traditional biological disciplines.[57] Conservation biology values biodiversity and calls for descriptions and interpretations of environmental systems in order to protect them. Ultimately, the success of this discipline in resolving environmental problems may lie in the ability of researchers, based in or outside zoos, to recognise the value-laden nature of their task and the practical challenges of applying research in management and policy settings. A growing circle of prominent scientists trained in conservation biology and related disciplines recognise the social and ideological dimensions of their work and appreciate that their awareness *can help* them make more insightful judgements about how appropriate their research methods and applications are.[58]

These and other like-minded people are making increasing calls for others in their community to recognise several issues.[59] First, practising conservation biology is grounded in advocacy for restoring ecological integrity to the biosphere. Second, all conservation efforts are contained in a 'human envelope'[60] made up of numerous people and organisations, therefore success will require attention to social and institutional processes. Third, current and future conservation scientists and wildlife managers will need to equip themselves with more interdisciplinary skills. Finally, the traditional conservation research community will need to move beyond the limited view of positivist paradigms and openly discuss their political, social and cultural commitments. The paradigms currently directing the work of zoo science essentially embody ideas and plans for shaping future relations between society, people and nature, and this 'social science' practised by zoo science needs to be opened up to qualitative inquiries in the social and policy sciences.*

Yet this discourse seems to be largely absent in the zoo community. There is an unmistakable preference in zoo research for positivist, biophysical inquiries. Furthermore, in Australia, there are currently virtually no social or policy scientists represented on zoo boards or in zoo research/conservation units. The outcome of this inclination is a partial view of and incomplete solutions for complex conservation problems (as opposed to treating symptoms). This is somewhat remiss given the high priority that zoos' formal strategy and policy documents place on issues such as using education and advocacy to reversing biodiversity declines. There is precious little social research—save for marketing studies—being conducted in these zoos to test the efficacy of zoo education programs (see Chapter 4). Some zoo administrators endorse what social science work is done, but it is largely the result of academic researchers' initiatives.[61] There is substantially more social research being conducted in American zoos, but a fair amount of this work does get caught up in positivist models: it is based in the quantitative schools of thought in psychology and seeks to predict the behaviour of zoo visitors.

If zoos wish to keep pace with advancing knowledge about how interdisciplinary work benefits environmental problem-solving, they may need to consider adjusting the social and institutional infrastructure that favours traditional research imperatives. Social and policy scientists could be recruited as members of zoo boards, scientific committees and research staff. These people could consider issues, such as the effectiveness of zoos' educational impact on society or could even turn the focus of inquiry to how zoos' own policy and decision-making processes affect their conservation role. This does not mean that zoos ought to abandon their biological research endeavours. Understanding the needs of captive animals and those being held for reintroduction programs is the responsible thing to do and is needed to inform management practices. Further-

* I am grateful to Dr Annie Dugdale for this final point

more, facilitating people's knowledge of and appreciation for biological and zoological aspects of nature is an important contribution to building an enlightened community. Nonetheless, biological research conducted in isolation from the broader social and political contexts that underpin such work will only ensure that zoos' contributions to biodiversity conservation remain incremental and self-serving. A more promising alternative is to integrate their biological studies with more qualitative social research. An undertaking like this, therefore, should draw on those qualitative social science methods which are capable of analysing contradictions between ideals and reality and addressing structural inequalities rather than merely seeking to maintain the status quo or to achieve greater efficiency of systems.[62]

Similar work has already been done in settings involving zoo staff, and resembles the exercises mentioned in Chapter 2. When the Eastern Barred Bandicoot was discovered to be close to extinction in 1991 after a management regime for the species had been in place for close to a decade, conservation professionals in Victoria used models developed in the policy sciences to undertake a detailed evaluation of the Recovery Program. The participants were conservation biologists, policy scientists, wildlife agency staff and zoo staff. Many had worked on the successful Helmeted Honeyeater Recovery Program (see Chapter 2). These people were essentially studying what the dominant research paradigm would typically *dismiss* as 'nonscientific variables'. The Recovery Team members critically examined their own organisational and professional values and processes to see what kind of impact these elements were having on efforts to recover the Bandicoot (for example, professional shortcomings, poorly defined timelines and responsibilities, negligible attention to social or economic issues, lack of effective leadership, poor communication and co-ordination among team members). Redressing these matters enabled the team to reverse the Program's poor performance and increase Bandicoot numbers in the wild. Since the mid-1990s, the Program's success has fluctuated. In the last several years, severe drought, competition in some conservation sites from kangaroos and predation by foxes has seen Bandicoot numbers decline. Additionally, vacillating political commitment, insufficient resources and ongoing personnel changes have weakened the problem-solving capacity of the Program, originally strengthened by members' consistent use of a policy sciences framework.

These issues demonstrate that evaluation mechanisms needed to be sustained if they are to increase a program's long-term effectiveness. Furthermore, critical inquiries in zoo settings could have applications that extend well beyond ensuring that zoos fulfil their research imperatives. Zoo research is part of zoos' educational imperative. Zoos can build on their current contribution to furthering their own and the broader community's understanding of environmental problems by increasing their capacity for conducting innovative research, disseminating those findings widely, and developing practical applications from such projects.

4

A Vessel of Higher Learning?

It is highly questionable whether the zoo should be out in the high-pressure entertainment market place touting for customers who otherwise would not be attracted by its traditional offerings—the zoo is and should remain a haven of education about and respect for animals, not a Disneyland or pinball arcade.

Editorial, *West Australian*[1]

Was the Ark of Noah's time a vessel of learning? We do not know the extent of Noah's (or his crew's) knowledge of the Ark animals and it seems unlikey that the animals served any specific educational function during their passage. It may be that the real instructional value in Noah's Ark is symbolic. When God passed judgement, sent the flood waters to scourge the Earth and then made a covenant with Noah, perhaps he had an apocryphal message.

In today's zoos, their education brief is supposed to be one a most important mission: that of facilitating public understanding of and concern for wildlife and broader environmental matters. Unlike Noah who did not have to bother himself with paying to maintain the Ark, today's zoos must maintain a regular and increasing revenue stream from their visitors. In return for visitors' paid entry, zoos must provide pleasant experiences. There may be some conflict between the imperative to educate the public about environmental problems and the need to bring in as many visitors as possible and ensure they have a good time. The way 'education' is understood and used in zoo settings can lend us some insight into this matter. That is, how do zoo professionals use education to achieve their goals, and what is it they are trying to say about environmental conservation? What challenges do zoo professionals face in their efforts to fulfil this goal? Exploring the nature of the educational experience at the modern zoo can help us to answer how well suited it is to being an institution of higher learning.

Education is a central policy goal for modern zoos. It is articulated in the mission statements of many institutions and in key policy and strategy documents. There are numerous international and national strategies which designate raising public awareness and action for biodiversity conservation as a high priority (see Appendix 2). Generally speaking, the formal aim of most modern zoos is to encourage an awareness and appreciation of animals and nonhuman nature. The fundamental stimulus for this 'education' process is meant to be the live animals that zoos possess, as is made clear by the World Zoo Conservation Strategy:

> living animals form the basis for education in zoos . . . the zoo visitor's susceptibility to educational information exists because of the attraction to the living animal, and animal collections are therefore the foundation of the enormous potential educational value of zoos.[2]

Both the presumed appeal of live animals and the collective reach of the world's zoos constitute their purported impact on the public 'consciousness'. Zoos remain popular institutions and are frequented collectively by millions of people each year. Their widespread appeal is seen by zoo professionals as a powerful opportunity to influence the views of many people in one place in a short time.

Zoo staff face several distinctive challenges in their efforts to deliver their education imperative to zoo visitors and the general public. These issues relate to the priority, scale, content and impact of their education programs. There are questions not just about *what* zoos should be telling the public, but about how well certain administrative arrangements support education goals and what impacts (if any) zoos have on the public's consciousness, given their expectations of what the zoo visit has to offer them.

Today, there are numerous competing beliefs about what should be the focus of zoos' educational imperatives as well. When the strong push for zoos to embrace a conservation ethic began in full in the middle of the twentieth century, zoo professionals started to ask whether zoo education had a larger purpose than to support school curricula and entertain the public. In the last twenty years, zoo educators and other commentators have been calling for 'the sleeping giant of environmental education' to be woken.[3]

Zoo educators' efforts to increase visitors' environmental awareness and action and to find the best means to achieve these goals are part of a broader debate about what are appropriate and effective messages and methods for environmental education. That is, should environmental educators merely encourage students to assimilate and reproduce factual knowledge in pursuit of a set of 'truths', or encourage students to become more politically astute and socially critical in relation to environmental issues?[4] Zoo education staff like Greg Hunt, former Assistant Director of Education at Melbourne Zoo, embrace the latter and believe that zoo programs should foster the development of:

- *knowledge* (of ecology, adaptations to environments, animal behaviour);
- *skills* (of observation, research, expression);
- *values* (appreciation, care, concern, empathy with other species, appreciation of effects of consumerism);
- *action* (participating in habitat and species conservation, models of appropriate behaviour in daily lives).[5]

The content of zoo education programs might restrict how well zoos are able to achieve these aims. Particular images and values are implicitly and explicitly imparted to the general and paying public through the presentation of captive animal collections. Some of the resulting messages may not be consistent with the aims of zoo professionals who are trying to engender positive attitudes to non-human nature and facilitate environmental advocacy. Administrative matters can also limit the delivery of environmental education in zoos. Zoo education services are primarily comprised of schools programs, which are funded and staffed through State education departments. Departmental priorities will not always conform to environmental education agendas. Moreover, the debate about suitable foci and processes for environmental education has yet to be resolved in primary, secondary and tertiary school settings, let alone in zoo education services.

Education in zoos can be broadly categorised by formal programs and the informal experiences of zoo visitors. Formal education schemes include school-based and community programs. Overall, program formats vary and cover a divergent range of topics. Most programs are oriented towards primary and secondary school levels. However, matriculation and preschool programs are available as well. A range of services are offered to schools which includes developing and providing written materials, conducting formal classes, developing curriculum, and providing in-service programs for teachers. Typically, teachers from public and private schools will organise class trips to the zoo to coincide with curriculum topics being covered in their own classrooms at that time. These excursions provide an opportunity for teachers to encourage children to apply concepts learned in school to the animals observed in the zoo, and these trips are thought to encourage children to have animal-related experiences not available elsewhere.[6] Learning can take place in a formal class taught by zoo education officers and/or through zoo tours conducted by school teachers. In the case of the latter, materials are often provided which orient teachers to the zoo and provide exercises for children to complete during their wanderings.

Public or community education is another component of zoo education services. These programs are offered to zoo visitors and the general community. They include adult education classes; special presentations by zoo staff (for example, behind the scenes tours), holiday programs (special classes), printed information, in-zoo guide services, maintaining public libraries, and information services.

Informal education in zoos is constituted by zoo visitors' sensory, cognitive and affective experiences of the zoo: 'non structured and non-obligatory learning'.[7] The overall atmosphere of the zoo, specific exhibit designs and any interactions a visitor may have with zoo personnel make up the major components of the interpretive environment of a zoo visit. Casual visitors—as well as children participating in formal schools programs or other groups formally visiting the zoo—are the primary audience for zoos' interpretive programs. In addition to these in-zoo experiences, zoo professionals use the public image of the zoo to deliver messages about the value of conservation and the zoo's role in these activities.

Generally speaking, those zoos with considerable financial resources are able to offer more fully developed education programs, although it must be said that funding education is not simply a matter of available resources. Administrative arrangements and program budgeting are political processes involving subjective decisions about the value of particular programs (see Chapters 5 and 6). Fluctuating policy and financial commitments to education are reflected in programs that vary in size, content and how integrated they are with the rest of a zoo's organisational structure. These variations result from traditional practices, professional priorities and personal initiatives. The degree to which education programs are integrated into the zoo structure is also influenced by the beliefs of staff members about the importance of the education function of zoos and how much influence education staff wield in their organisations.

Education programs have traditionally been the responsibility of zoos' education services, and were, until relatively recently restricted to school programs. Adelaide Zoo, Perth Zoo, the Territory Wildlife Park, Taronga Zoo, the Western Plains Zoo, Melbourne Zoo, Werribee Zoo and Auckland Zoo conduct these services in cooperation with, and through the assistance of State and Catholic Education Departments' extension services. These divisions provide ancillary services by seconding trained education professionals to institutions such as zoos, museums and art galleries. Co-operative arrangements between zoos and State Education Departments will often entail the provision of infrastructure support from zoos along with other cost-sharing mechanisms. For example, at the Western Plains Zoo facilities for the Education Service have been provided by the Zoo, as has 52 per cent of the administrative costs. The NSW Education Department has carried the remaining 48 per cent.

The delivery of *environmental* education can be limited in part by two or more institutions managing zoo education services. The predominance of seconded teachers in zoos can result in a 'two bosses' syndrome where teachers must try and service the priorities of the zoo *and* the Education Departments.[8] The State Education Departments' priorities are for zoo education to service its curriculum and to provide teacher training. Yet in the zoo industry, there is awareness that zoo education services ought to incorporate a vision of education that is broader than a focus on school programs and which, in some cases, prioritises environmental education. Trying to realise all these aims may test an education officer's priorities.

While State Education Department curriculum will include some environmental education, there will be other subjects that a zoo education service will be asked to provide (such as biology, zoology). Furthermore, State Education Departments do not have an imperative to provide environmental education to the general population of zoo visitors or to the broader community. Zoo education officers may find that they are having to exercise some caution about the time they devote to activities or programs that fall outside their duties as defined by State Education Departments. In these cases, an almost exclusive emphasis on school visitors results. Furthermore, as the average length of secondments to zoos has been three years, a regular turnover of staff predominates in zoo education services. This constant change can create instability in these departments' policies and minimises the ability of such staff to make regular and continuous contributions to overall zoo policy. Some zoo teachers exercise a high level of personal initiative in attempting to further integrate the service into other zoo programs and in lobbying for policy changes. While the motivation of a few individuals can make significant positive impacts on the success of a project, a particular zoo's 'real' commitment to education will be revealed by whether zoo professionals have devised formal policies and institutional structures that foster the development of fully-funded and extensively-integrated education programs.

The traditional organisational and conceptual isolation of the education function in zoos has come under close scrutiny of late by those who feel that zoos might be paying lip service to their education role. Some of those critics are zoo professionals who have been eager to close the gap between education rhetoric and actual practice. Until recently, the Zoological Parks and Gardens Board of

Victoria (ZBV) and Perth Zoo had highly developed education services which wielded significant influence on zoo policy. Education staff were able to design and deliver innovative programs and actively participate in determining organisation-wide priorities. They were able to do these things in part because both zoos exemplified the trend towards having zoo education staff be part of the executive management team. For example, the Director of Education at Melbourne Zoo was a senior manager with responsibility for overseeing the education services of the three ZBV properties (Melbourne Zoo, Healesville Sanctuary, Werribee Zoo)(see Chapter 5). While this structure did reflect a move towards greater centralisation, it also helped to provide greater representation of the Education Service's needs and concerns in key decision-making fora. Around that time, the ZBV also developed a Zoo Education Strategy:

> [The Strategy] is one that provides high quality school education programs to meet the needs of a wide range of students and levels of schooling; is one that realises the need to develop an understanding in a range of VIP's of our role and function; and is one that caters to the interests of its visitors through the provision of high quality public education programs. It is an approach that embraces the whole zoo community and its visitors through the provision of high quality public education programs. It is an approach that embraces the whole zoo community and its visitors.[9]

The Strategy considered that education services ought to be the responsibility of the entire zoo and sought greater integration of the education function and department into the policies and functions of the Zoo. This 'whole-zoo approach' was meant to foster clearer understandings and a better appreciation among all zoo professionals—from animal keepers to marketing managers—of education's importance for zoos' conservation role. Unfortunately, after a significant organisational restructure in 1995, the Strategy never resurfaced. Given the failure of the Zoo Board to endorse and implement the Strategy, it seems questionable whether education imperatives will receive any more attention than they have to date.

Perth Zoo's administration of education programs illustrates a similar trend. The Education Service is operated entirely by the Zoo and, until recently, the Education Manager functioned at a senior management level. These arrangements had helped ensure that education issues were heard in decision-making forums, and facilitated a greater degree of operational flexibility. Education staff were relatively free to vary their programming and offered, in addition to the schools program, community education programs, a library service and others. In response to a decrease in government funds and an increase in the number of school students attending programs, the Service adopted an entrepreneurial approach to raising revenue. It secured a substantial amount of funding for its operations from outside the Zoo, earning in excess of $200 000 over an eighteen-month period in special purpose grants from the public and private sectors. Like their Victorian counterparts, however, the Education Service was downsized and subsumed by the Visitor Services program in 1995.

Story-telling on the Ark

The subject matter and administration of education programs in zoos influence how well zoo professionals achieve their goals to enlighten zoo visitors and the broader community. What other relevant considerations are there? Two critically important factors influencing the effectiveness of contemporary zoo education are people's motivations for visiting zoos and the kind of informal learning experiences zoos engender.

There is little doubt that zoos are a popular institution. Some behavioural theorists suggest that zoos are similar to other leisure-based informal settings, where factors of relaxation, escape from stress and tranquillity are valuable experiences in a largely high-stress society.[10] Zoos present 'nature' to people in what is, and has largely been an urban recreational setting. Elements of a zoo visit target providing an escape from the city, and are meant to facilitate an enjoyable—if not readily apparent—learning experience. The theory goes that visitors may not necessarily realise they are being educated by their visit. Indeed, visitors may need to be persuaded

This Asiatic lion exhibit is typical of Western Plains Zoo, Dubbo, where moats separate visitors from apparently free-ranging animals. These exhibit techniques, developed by Carl Hagenbeck in the early 1900s, signalled an interest in more realistic viewing and better conditions for animals.

A child views the mandrills at Melbourne Zoo. Recent use of plate glass and naturalistic exhibits allows visitors a sense of being nearly inside an animal's 'habitat', where animals are more inclined to engage in 'wild' behaviours.

Even before visitors enter Taronga Zoo, they are greeted by two contrasting images of the zoo. The front gates symbolise the colonial origins of the major Australian zoos. The platypus sculpture signifies the more modern role of the zoo—conservation.

A female Black Rhino at Western Plains Zoo, Dubbo. Will modern science 'save' the black rhino? As with other similarly managed species, its long-term conservation status is far more ambiguous than zoos' public relations machines would have us believe.

into their education. Studies suggest that most zoo visitors are primarily seeking a safe, attractive and entertaining social experience, rather than a more serious learning encounter.[11] Moreover, many professionals in the zoo community purposefully present the zoo experience as recreational and enjoyable, in order to attract people. One zoo horticulturalist and education specialist described the attractive exhibits and grounds in zoos as the 'aesthetic carrot that lures them towards a unique learning experience'.[12]

These 'beautiful surroundings, ecological integrity and natural behaviour of the animals' exhibited in the zoos'[13] are supposed to trigger this unstructured and voluntary learning experience. Inspired by zoo professionals' interest in animal welfare and environmental education, naturalistic habitats are supposed to replicate the aesthetic and/or functional experience of a species' wild habitat. Modern naturalistic designs are meant to be a popular catalyst for creating a stimulating environment for people as well as captive animals.

Today, most Western zoo professionals believe that the sight of caged animals is a self-defeating and inappropriate strategy for zoos, one that does little to engender respect for the institution or appreciation for the wildlife kept in it. Instead, contemporary exhibits are designed to provide better care for animals and to provide a stimulus for visitors to make the connection between the animals they see in simulated habitats and what exists 'out there'. The 'naturalness' of exhibits is supposed to meet the psychological and physiological needs of the animals, thereby increasing the incidence of behaviour one might see if viewing these animal 'in the wild'. Studies of zoo visitors have found that people are more attracted to where animals were active and where there was a sense of realism to the exhibit.[14] Hence, zoo professionals assume that observing animals that are more active and engaged in natural patterns of behaviour in their naturalistic enclosures may facilitate a more enjoyable and educational experience for zoo visitors, and trigger in them a greater respect for 'wild' animals.

A prominent zoo architect, John Coe, has suggested that in order for zoo visitors to learn more they must be exposed to more interactive environments than these zoos have offered to date. He was instrumental in developing landscape immersion exhibits in zoos. These designs are meant to take the experience of viewing animals in naturalistic habitats one step further by placing zoo visitors 'inside' an animal exhibit. Surrounded by the same 'natural landscape and attendant multi sensory environment clues as are the animals . . . the visitor is expected to feel as if they left the zoo and entered the [animal's natural habitat]'.[15] The visitor is encouraged to step into the animals' place, to be aware of their fear about leaving their own environment, and to develop a deeper understanding and appreciation of the animals they are viewing.

Grouping several animal exhibits by themes is another method zoo professionals employ to convey environmental messages to zoo visitors. David Hancocks, another prominent zoo architect and director, identified several different typologies for zoo exhibits; however, these are not mutually exclusive as most zoos have been planned to include a selective combination of the five themes.[16] *Systematic* displays present groups of zoologically related species in close proximity to one another and are reminiscent of zoos from the nineteenth and early twentieth centuries. These designs are meant to confer educational benefits to zoo visitors by relating different species according to their morphological characteristics, but understanding how species interact with their environments is not a central focus.[17] *Zoogeographic* collections are organised according to animals' continents of origin. Unlike the systematic pattern, these designs do not necessarily separate species. Rather, different species are linked together on the basis of their geographical distribution. *Habitat* displays group together species from similar habitats such as rainforests, savannas or aquatic environments. This design trend is growing in popularity and is most consistent with ecological issues in environmental education. The *Popularity* of certain species is another exhibit philosophy. Charismatic fauna may often be strategically placed in central locations in order to attract the attention of zoo visitors. Finally, the *Behavioural* exhibit philosophy informs animal collections on the basis of ethological factors such as swimming, burrowing or flying. Rather than focusing on species *per se*, the primary interest of this approach lies in illustrating relationships between physical characteristics, psychological-social characteristics and adaptations to the environment.

Most zoos mix different patterns of display. A typical exhibit arrangement includes some mix of systematic, zoogeographic, habitat, popular and behavioural themes. For example, the Melbourne Zoo's Master Plan mixes the tenets of habitat and zoogeographic themes by designating several bio-climatic habitat zones such as the Tropical Rainforest, Eucalypt Woodland or Savanna. Incorporated into each zone are geographic subdivisions: African, Asian, Australian and South American. The main thrust of the Plan 'is about habitat recreation, to display animals in naturalistic exhibits, demonstrating the links between both plant and animal species existing within the one habitat . . . to break down the taxonomic groupings on which the Zoo was previously developed'.[18] Zoo staff hope these new ways of presenting animals to visitors will promote important environmental values such as the uniqueness and interconnectedness of diverse forms of life.

The older metropolitan zoos in Perth, Adelaide, Melbourne, Sydney and Auckland provide the most striking examples of the old and new in both their collection philosophies and exhibit designs. While these zoos often contain collections of exotic and native species more diversified than the wildlife parks and sanctuaries specialising in native fauna, a necessary consequence of this variation is that it encompasses more inconsistencies among exhibit philosophies. Components of these zoos represent traditional, conservative systematic and popularity collection philosophies, while newer exhibits and master plans embrace zoogeographic, habitat and behavioural groupings. This contrariety is an impressive dynamic. One can hardly avoid noticing the old and the new existing side-by-side, as well as efforts to make the old function and look like the new, all evidence of the staggered nature of development given limited resources. Intermittent modernisation of zoo exhibits most likely frustrates zoo professionals' efforts to convey progressive conservation messages to the general public.

Another important basis of comparison for exhibit styles can be found between the metropolitan zoos and the wildlife park/

sanctuary formats. Generally, the wildlife parks, sanctuaries and open-range zoos are more successful in portraying a 'natural' or 'wild' setting than the metropolitan zoos. In the cities, zoo professionals are constrained by their urban surrounds, and produce simulations of nature that tend to be more contrived than those habitats constructed in less developed areas. The rural and more natural atmosphere of the wildlife parks and sanctuaries, in conjunction with collections that are primarily comprised of native species, may convey enhanced experiences of habitats (and conservation information) that surpass those of metropolitan properties.[19]

To highlight interdependence between animals, ecosystems and people, zoo staff can also use horticultural planning in city or country zoos. Through the selection, use and maintenance of appropriate plants, a zoo horticulture department can contrive provocative naturalistic landscapes that may heighten visitor awareness of animal/habitat relationships. The vision is that zoos can become environment parks with a strong botanical orientation that integrate facets of botany and zoology in ways that simulate nature.[20] John Robinson, a prominent member of the international zoo community, believes that the zoo must surpass its botanic and zoological roles and become a 'biopark':

> [A biopark] combines the attractiveness of living plants and animals with exhibits that explain their structure, physiology, history and interconnectedness. It is composed of subject matter from existing institutions, such as museums of natural history, anthropology, art; botanical gardens and arboreta; and zoos and aquariums.[21]

More explicit means of convening the kinds of issues Robinson discusses are occasionally found in signs and graphics.

In addition to naturalistic habitats, signs and educational graphics constitute what zoo professionals hope is a significant component of zoo visitors' interpretive experiences. Graphics include signs with written text plus art. Like informal education techniques used in museums and national parks, the design, function and educational philosophies of zoo signs vary. Certainly, there has been a shift away from the traditional small 'name and distribution' signs providing the basic animal identification. When zoo educators and interpretation professionals realised they could be telling visitors more, there was a noticeable shift towards the production of signs with colourful illustrations, large print texts and interactive components. Signs and graphics are designed to attract and hold visitors' attention and engage their concentration long enough for visitors to read them and have positive reactions to the content.[22] Attitudinal and behavioural changes towards the environment among visitors are considered the ideal outcome by many in the zoo community.

Although signs are the most obvious and, some would say, important communication devices available to zoos, it remains that they are not particularly well understood or evaluated. Several studies have raised doubts about the frequency and duration of sign reading. It is believed that only a minority of visitors read the signs and spends minimal time doing so.[23] Nonetheless, some visitors report that when they have questions about the animals they are viewing they first look to the signs at the front of exhibits for information. Similarly, a majority of visitors surveyed at Australasian zoos agreed that they would be disappointed if an exhibit did not have an interpretive sign and that they enjoyed learning from that information.

Irrespective of whether visitors view animals in naturalistic, landscape or immersion exhibits, the degree to which zoos can succeed in encouraging visitors to appreciate the zoological, ethological and ecological aspects of animals remains unclear. The nature of zoos' animal collections, the manner in which the collection is displayed, and how the public interprets that collection are all relevant matters. The potential educational impact of a day at the zoo may be influenced by zoo visitors' expectations of a fun day out and the degree to which zoo professionals help to maintain—rather than challenge—those expectations.

In their definitive social and cultural interpretation of the zoo, *Zoo Culture*, Bob Mullan and Garry Marvin provide further insight into the origins of visitors' recreational motives by comparing the experience of visiting museums and galleries to that of zoos. They suggested that visitors to museums and galleries revere the culture that is represented there, despite the fact that they are not always clear how they should interpret it. Mullan and Marvin suspect that for zoo visitors there is neither the need nor the motivation to understand in any great depth the animals they are viewing. They believe that the popularity of the zoo lies in the fact that it does not intimidate people. Visitors can enjoy themselves without having to possess a lot of knowledge about the animals they are viewing.

Considerations of how people relate to the animals once they are there seem to support why zoos are so popular. Several studies have found that visitors seem more concerned with the characteristics, behaviour or welfare of individual animals rather than intellectualising on complex details involved in say, species conservation.[24] One researcher found that, even though visitors prefer exhibits that satisfy both themselves and the animal exhibited, their preference was informed more by an affection for animals, rather than an awareness of animal ecology.[25] She also observed that visitors failed to gain a better understanding of wildlife, despite a higher incidence and degree of exhibit 'naturalness'.

Visitors are also thought to have a relatively short span of attention. While zoo architects, such as John Coe, assert that visitors will have a richer experience by virtue of having to find animals camouflaged by their naturalistic enclosures, others have uncovered evidence to the contrary. Stephen Bitgood and Arlene Benefield found that where animals could be readily observed being active, they held the attention of visitors longer.[26] Visitors have been known to become frustrated or lose interest if they must work too hard to find the animal on display. For example, at Melbourne Zoo the Sumatran Tiger exhibit features a simulated rainforest habitat, replete with dense vegetation. Sometimes the vegetation hides the tigers, and visitors can be observed moving on quickly to other exhibits if they cannot immediately see the tigers.

Another appealing aspect of a zoo visit has been the sense of being close to 'wild' animals. Mullan and Marvin suggest that the sense of excitement people gain from viewing zoo displays is directly linked to the animals' symbolic value. The animals represent something hidden or distant from the experience of our daily lives. A case in point is when animals are presented or interpreted as 'dangerous'. Mullan and Marvin suggested that we project this characteristic onto animals. An animal is often considered 'dangerous' because of its potential to harm, even kill, *people*. Yet, it is only when humans disturb an animal's (simulated or natural) habitat that a potentially threatening situation is created.

Mullan and Marvin suggest that because the presentation of an animal in captivity does not allow for its full repertoire of 'natural' behaviours, visitors are more inclined to project certain cultural interpretations onto it. Zoos may encourage people to spend a lot of time thinking about *themselves* in relation to an animal rather than about the animal itself. At Taronga Zoo, a life-size, two-dimensional cut-out of a Kodiak bear is featured outside the bears' enclosure. Next to this silouhette is a measure where visitors can stand and compare their height with the truly impressive stature of the bears. Is the implicit message in this exercise the potential danger that a life-size bear represents to a person? If people are encouraged to feel at all threatened by viewing 'dangerous' carnivores they might also be encouraged to feel a sense of alienation from non-human nature. Yet this is the very attitude that zoo professionals wish to discourage.

Other factors make the experience of viewing an animal in the zoo qualitatively different from seeing them in their natural setting. Mullan and Marvin spoke about the different ways of naming and representing animals. Some zoos still give animals human names. This form of anthropomorphism can highlight an animal's individual personality more than featuring what its uniqueness as an animal or species. Assigning scientific names to zoo animals has a more educative bent to it, nonetheless, it remains an effort to classify and order non-human nature according to a particular cultural scheme. Moreover, it is unclear whether visitors take note of these names and, when they do, how they make use of them.

A preoccupation with individual animals might be problematic for zoos' (environmental) education imperative. Where there are more *human* constructions of animals or *human* characteristics projected onto animals than there are understandings of animals as independent entities (with rights of their own), then zoos risk sending mixed and conflicting messages. These confused meanings might limit the degree to which zoos can attend to some animals' physiological and psychological needs, deliver sound lessons to zoo visitors about respectful relationships with non-human nature or be forthcoming about the nature of its conservation programs.

Promoting the Ark or Educating the (Human) Passengers?

The promotion of zoos can also convey ambiguous messages to the public and compromise education goals. Why do zoos, however unknowingly, propagate these illusions? Part of the answer may lie in the persistent tradition of the zoo as a popular recreational institution. Mullan and Marvin suggested in their analysis of the zoo that this institution had never achieved the same cultural status as art galleries and natural history museums. They asserted that once animal collections became accessible to the public, the animals and the institution in which they were displayed, seemed to lose their status. Unlike the items held by art galleries or museums, living animals were not valued intrinsically, were not considered unique, nor did they symbolise significant historic or cultural processes. Zoos' education imperative was as basic as placing simple labels on enclosures to identify animals. Moreover, people were not necessarily coming to the zoo in search of enlightenment. Indeed, we have already seen that some of the early zoo founders were loathe to have their institutions encouraging the 'unlearned' to stare at exotic beasts in a somewhat stupefied state. Nonetheless, unlike museums and galleries, zoos have not been able to command the same kind of respect in which to these élite cultural institutions are held. Perhaps zoo professionals' pursuit of a role in conservation and education is partly a search for a legitimacy and status rivalling that of other cultural and educational institutions.

In addition to a quest for validity, the more practical concerns with a zoo's commercial imperatives may create a push to promote a positive image to government officials and the corporate and general community. Zoos' financial viability rests quite heavily on government grants and revenues generated from their gate takings and from corporate sponsorship of special events and capital works developments. In order to ensure the continued flow of these monies, zoo professionals work to ensure that government officials, business leaders, zoo visitors, and to a lesser degree the general community, are pleased with the zoos' policies and programs.

It is primarily the role of marketing managers, public relations officers and media co-ordinators to develop and maintain a favourable image and highly visible profile for their institution, although CEOs, directors and other very senior managers may occasionally participate in these activities. These staff are typically quite sincere in their devotion to and pride in their institutions: they occupy important places in a zoo's organisational arrangements and are therefore empowered to define—explicitly and implicitly—education policy. Moreover, these staff often have different views from their colleagues in zoo education departments about what 'education' in zoos is and/or ought to be and what are appropriate ways to promote the zoo. They are not likely to have professional expertise in learning theory, practical experience delivering effective education programs or an appreciation of how institutions influence public opinion, nor to recognise how their actions contribute to a substitution of means for ends.

Promotion occurs in zoos' annual reports and other official policy documents. The media is also a particularly important source of image-creation and is used quite extensively by most zoos to encourage the public to view them in as favourable a light as possible. Successful endangered breeding programs, exhibit openings and various other 'good news' stories are regularly fed to the media. 'Good' publicity results when these achievements are highlighted and this information is widely disseminated. It is very likely people's awareness of the changing role of zoos has resulted from the intense effort by the international zoo community to publicise that shift in purpose. However, the media cannot be relied upon *not* to bite the hand that feeds it. 'Bad' publicity for zoos are stories or reports that generate any controversy and portray a zoo in any negative light. For example, when the Australian Broadcasting Corporation aired the final episode of its documentary, *Zoos Company*, in which a zebra had to be euthanased because of problems transporting it interstate, many zoo staff considered this to be very damaging to zoos' reputation. Bad news can be considered to be *better* news by press and television journalists who are not averse to preying on zoos' misfortune when something goes awry, or when certain groups wish to direct the public eye towards zoos' more unpalatable practices. Animal welfare lobbyists' could not have advanced their causes against zoos without calling upon the assistance of the media. Consequently, zoos' media managers or public relations officers work quite hard to develop working relationships with a range of media contacts. In this way, they increase their chances of maximising opportunities for good publicity, minimising damaging publicity or, at the very least, ensuring they have a chance to respond to negative stories.

When zoo visitors and the general community hear about 'good news' zoo programs, in many cases, they are being asked to believe that zoos are 'saving' species. Newspaper articles, publicity materials and exhibit signage are some of the more common vehicles zoos use to present information about their contribution to species restoration programs. Similarly, an information brochure for the Animal Gene Storage Resource Centre of Australia at Taronga Zoo reads, 'The future: conservation—saving our animals'. A sign in front of the Gold Lion Tamarin exhibit at Taronga proclaims: 'Tamarin saved! Numbers increased and future assured'. In the Weekend Section of the Melbourne *Age*, readers are told, 'you can sponsor your choice of wild animal . . . at the Melbourne Zoo . . . from $50, you can help save endangered or vulnerable species'. A local Landcare group planted one section of the Perth Zoo's African exhibit. The narrative on the sign in that area begins with 'Saving the Savannah . . .'. A small article in a Missouri paper, the *St Louis Post Dispatch*, claims that 'With biodiversity behind bars, zoos can preserve endangered species and educate most people'. The *Sydney Morning Herald* ran an article with the title 'Zoos in the business of protecting wildlife'.

What do zoo professionals mean when they refer to a species or animal being 'saved'? Modern zoos have justified the keeping of (threatened) animals because these creatures can serve as ambassadors for their wild counterparts. The value of a zoo-based research or species management program for biological conservation is how those activities can assist in reintroducing threatened species to their *natural* environments. It is not always clear in promotional materials how and when—if at all—this might happen. In many cases, there is no habitat to reintroduce captive animals into, or the habitat has yet to be secured so that some species will remain in the zoo indefinitely. Mullan and Marvin ask whether these animals shouldn't be considered some sort of refugees, rescued from what was once their home but now maybe relegated to a permanent state of limbo. What does the visiting and general public, or potential and current corporate sponsors understand about the status of these populations when they are portrayed as being rescued by zoos?

Furthermore, the intense publicity afforded to high-profile captive breeding programs for charismatic vertebrates could distract public attention away from the importance of habitat preservation, the value of lesser known vertebrate and invertebrate species and—perhaps most significantly—the complexity of solving the broader loss of biodiversity. In so doing, the justifications of captive breeding may speak more to human interests than the interests of animals—even where species are 'saved' from extinction. That is, one outcome of intense publicity efforts can be a lack of clarity about whether zoos are more effective at promoting *their role* in conservation than they are at promoting conservation itself. These are two different things.

Other messages can be equally misleading. When visitors to Australasian zoos were surveyed, most of their negative comments highlighted the cramped conditions of some enclosures, particularly those housing the larger vertebrates such as elephants, giraffe and lions.[27] The data suggest that some visitors may have experienced a level of discomfort when viewing these animals since they tended to describe their feelings as 'sadness'. While the unfavourable comments intimate that many respondents tend to have adverse reactions to seeing animals in cages, some appear to be reassured by the naturalistic designs which are now so prevalent in modern zoo exhibits. Certainly the open-range zoos and more expansive wildlife parks and sanctuaries evoke commendatory responses from visitors who appear to believe animals are happier in 'natural', more open environments. Notwithstanding the fact that people can only imagine (not directly experience) what an animal senses, a simulated zoo habitat (no matter how realistic it may be) is not the same thing as 'the wild'. A person will be projecting their experience of the animal's life in captivity (albeit in greatly improved conditions) onto that animal, not its life in the wild. Moreover, several visitors' use of the term 'natural' and 'habitats' when referring to animal enclosures imply that some respondents are not making a distinction between 'real' and simulated nature. When zoos use the words 'natural' and 'habitat' in their exhibit signs and other materials to refer to animals' enclosures, they risk creating and reinforcing visitors' misunderstandings.

The zoo's capacity to deliver sound conservation or environmental education will also be influenced by the way zoo professionals work to ensure that the zoo is a commercially viable, recreational institution. There are debates about the impact of commercial practices on education principles and practices. Many keepers, curators and educators are concerned that 'the role of the zoo is being circumscribed by the perception that people want a fun day out.' There is a tangible sentiment among some zoo staff that zoos are in 'show business, selling recreation not conservation'. Other staff, often CEOs, directors, senior marketing/visitor services managers acknowledge these opposing values in different ways. Some marketing and media staff are quite concerned about 'tacky' messages which might jeopardise zoos' educational goals. These staff admit 'that it's really hard . . . to determine what is overstepping the mark'. They 'agonise over it' and regularly consult with animal management staff to try and find 'the fine line'. Generally speaking, there is widespread acknowledgment among most zoo staff of some level of conflict. As one senior marketing manager said:

> there is always a conflict between the marketeers who are responsible for getting in the visitors and increasing the yield per customer . . . and the keeping and conservation staff who are concerned that we are overcommercialising the product.

However, the potential value of this awareness is compromised when staff argue that 'commercial objectives do not threaten conservation'. These staff recognise ideological discord, but they shift the argument away from the problem of conflicting values to giving primacy to the commercial imperatives when they point out that 'if you ignore the commercial stuff you are really putting [the zoo] at risk'. They assert that 'we are only free to do those things we have the money for'; that without the revenue, 'the zoo does not have any conservation imperatives'. A customer-service orientation then predominates: 'we can't take people's money without delivering something' as the 'reality is that we need to attract visitors' and 'our primary aim is to attract customers and [provide for] customer satisfaction'.

This emphasis on maintaining, increasing and satisfying visitors, in turn influences the question about what zoos do or should deliver. Many zoo professionals agree with the academic research, which suggests that people are motivated to come to the zoo for a leisurely day out—as opposed to enlightenment. They also see the zoo as having to compete for visitors' leisure dollars that might otherwise be spent on some other, more 'exciting' venue. The way this information is exploited, however, depends on staff members' particular views about zoos' education role. Some staff feel that the zoo should actively influence its public, irrespective what motivates them to visit or support the zoo. Two curators spoke of an ideological split in zoos. The first identified 'two schools of thought in the industry': to 'give the people what they want' or to 'direct people to what they want'. As the second put it, 'There are two camps in zoos . . . those who think the public is not blind to conservation and those who say we must have lions, tigers, and giraffes . . . [it] creates difficulties in getting along the path'.

It seems that those who are most concerned about maintaining an optimal income level for the zoo are wary of loading zoo messages with too much seriousness about conservation or environmental matters. Assuming that visitors *are* less interested in learning about what they can do for conservation than they are in having 'an exciting adventure', these zoo staff believe zoos should avoid risky strategies like overtly trying to change public opinion. Rather, zoos should be able to compete with other recreational venues by 'building new exhibits that are [more] exciting . . . and advertise [the zoo] as an exciting place to visit'. This argument is gaining a foothold in zoo policy of late, and it construes 'conservation' or 'environmental' education and its attendant messages as negative, overbearing, depressing or just not inherently 'fun'.

Therefore, if some people believe that these programs might discourage people from coming to the zoo, and they worry about visitor revenues declining as a result, it seems reasonable to assume that extensive conservation or educational programming might be viewed with scepticism or worse, be discouraged. Those staff or other people who are concerned that conservation and education are being compromised by commercial activities are typically dismissed as 'purists' and labelled naive because they fail to appreciate that 'people can't do education, research and conservation if the money doesn't come in.'

Is it possible to determine that zoo visitors are, in the first instance, *not* interested in conservation and, in the second, going to visit the zoo less often if *more* conservation education programs were implemented? It is not clear. In Australasian zoos, there is very little formal evaluation of visitor behaviour. Given the increasing importance being placed on increasing the number of zoo visitors and the money they spend while in the zoo, most research is marketing-based. This quantitative type of evaluation is based on visitors' opinion and focuses primarily on how satisfied they are with their zoo experience and with zoos' overall performance, and when they are likely to visit again. Sometimes questions are put to visitors in a scale format, asking them to indicate a level of agreement or disagreement with whether the zoo fulfils its conservation obligations sufficiently. In my survey of 1600 zoo visitors, I found most of them to be supportive of zoos. It does not seem surprising then that zoo-based marketing surveys would find that a majority of visitors 'agree' that the zoo does a good job in 'conservation'. What this line of inquiry does not tell us, however, is what visitors' environmental values, knowledge or behaviours are, nor does it examine what *impact* the zoos' formal and informal education might have had on them. Operating under significant time and resource constraints, education service staff do try to obtain, at the very least, informal feedback on their programs from teachers bringing their classes to the zoo.

In Australasia, the academic community has tended to do most of the qualitative research on visitors and environmental learning. In my research I asked zoo visitors what kind of information they had learned about or were reminded of as a result of their zoo visit. A small percentage of the respondents certainly had some pre-existing awareness of the concepts of biodiversity, adaptation, habitat and certain environmental problems and reported that they had learned about some of these matters through their zoo visit. What was readily apparent was respondents' appreciation of these and other values and of the need for conservation, as was their recognition that environmental problems have human causes, which all levels of society are beholden to address.

Should zoo education do more than affirm existing knowledge and values of its visitors? Where visitors (or the public at large) are uninterested or even hostile to environmental issues, should the zoo back away from shifting public opinion? The survey data from the Australian zoo survey suggests that most zoo visitors have largely generalised knowledge about environmental conservation and demonstrate low levels of activism. This suggests that more work needs to be done to increase their understanding of contemporary environmental dilemmas and their support for environmental reform. It may also be that because most visitors do *not* have strong ties to conservation groups, and showed *less* interest in learning about conservation activism relative to other topics, zoos' programs are not effective in attracting conservation-minded people or facilitating more 'environmentally-conscious' behaviours in existing visitors. I found that those visitors with environmental affiliations and higher levels of education were more discerning about zoos' conservation policies and *supported* the notion that zoos should promote environmental advocacy.

Australia's poor track record in the loss of biodiversity and a host of other environmental problems suggests quite a strong imperative for educating the public. Certainly, official mandates such as the *National Strategy for the Conservation of Australia's Biological Diversity* implore all institutions to inform the public about conservation and to help to motivate them to make active contributions to biodiversity conservation. What this and other policy and strategy documents like it do not offer, however, are insights about the symbolic value of institutional behaviour, specific guidelines for facilitating individuals' behavioural change, and what priority education activities should take relative to other institutional imperatives.

When zoo professionals construe 'conservation' or 'education' as something negative or dreary, they tend to revert to the more traditional idea of zoos as places of recreational entertainment. In turn, zoos increasingly rely on commercial activities to cure their financial woes. A number of strategies that are currently being used in zoos to boost flagging revenues and combat rising operating costs tend to focus, not just on increasing the number of visitors to the zoo, but on getting those people to spend more money (increasing visitor yield) once they are inside. For example, anyone who has been to a zoo would be familiar with the zoo shop or restaurant. A zoo shop typically sells items such as plastic animals, key chains, jewelry, T-shirts and windcheaters with a zoo's name, stuffed toys, books, postcards and other souvenirs. Zoo cafés typically feature a selection of fast food and vending machines enable visitors to purchase soft drinks and candy. The shops and machines are often very close to the entry and exit points of the zoo in order to increase the chance that visitors will buy something.

While these commercial outlets do generate much-needed revenues, what do they tell zoo visitors about environmentally responsible behaviours? People in modern society increasingly calculate 'success' by how much they consume. Barring all the evidence, which shows that the accumulation of material wealth does not equate with personal happiness or equitable societies, rising levels of consumption require that considerable amounts of resources be exploited. In turn, various pollutants and poisons are created and whole ecosystems are degraded beyond repair. On one hand, the zoo strives to teach people about how humans have caused the plight of animal species and the need to use resources wisely. Yet, in nearly the same breath, it is hoped that visitors can be cajoled into consuming ever-increasing amounts of material goods. Zoo shops and restaurants are rarely used to discourage these tacit messages. While a host of educational books on Australian and threatened wildlife can be found, environmentally-friendly products or en-

This sign at Taronga Zoo typifies grandiose statements about zoos' contribution to international tiger conservation efforts. Do zoos intend visitors to equate a colony of captive tigers with securing their existence in the wild?

A goanna in the Northern Territory's Wildlife Park. Encountering free-ranging animals may heighten visitors' sensation of being 'out bush'.

Signs, like these at Adelaide and Perth Zoos, enable corporate sponsors to display their putative interest in 'wildlife' and conservation.

Progressive, ecocentric values are embodied in the native grasslands exhibit at Melbourne Zoo, which are intended to raise visitors' awareness and appreciation not only of wildlife but also of local and regional ecosystems.

vironmental education materials are not prominent features of either the zoo shops or zoo eateries. There is considerable potential for these venues to do more. In order to support themselves, many environmental NGOs such as Greenpeace, the Australian Conservation Foundation or the Wilderness Society, sell 'green' products and educational materials through mail order catalogues or their own shops. Zoos could help promote these organisations and environmental education by featuring at least some of the NGOs' products in zoo shops. Yet this policy is virtually unheard of in Australasian zoos.

Zoo professionals may compromise their institution's potential to facilitate environmental awareness and advocacy when they overemphasise commercial principles and activities. The promotion of unsustainable behaviours suggests a low level of awareness among some zoo decision-makers about the principles of *ecological* sustainability, and about how institutions can shape public opinion. It also creates another reason to reform zoo policy so that it supports, not conflicts with ecological values. If some zoo staff are eventually successful in convincing zoo administrators and other relevant policy-makers that environmental education is an activity that requires increasing amounts of support, the next task will be to ensure that the zoo is better equipped to facilitate environmentally-responsible behaviour.

Some psychologists/researchers assert that measures for helping society shift to a more ecologically sustainable lifestyle will be more effective if educators and conservation advocates aim to change environmentally relevant *behaviours* rather than *attitudes*.[28] Much behaviour will have to change in a relatively short period of time and stay changed if ecological sustainability is to be achieved. Certainly there are myriad variables that interact and determine people's lifestyles and actions. Moreover, it can be difficult to determine which factors have the most formative effects on the development of environmentally responsible behaviour.[29] While zoos cannot realistically be expected to address all forms of environmentally irresponsible behaviours, they might at least be able to take a more proactive and critical approach to environmental education.

Questions about the effectiveness and appropriate focus for zoo education programs are embodied in criticisms about zoos that come from outside *and inside* the zoo community. The zoo community often is labelled as overly conservative when some of its members promote the notion that raising environmental awareness can and should be constituted by passive, apolitical learning. When zoos assume this traditional stance, they help to marginalise environmental interests and miss the opportunity to educate their visitors about progressive environmental knowledge and values. This quandary reflects the broader debate about whether environmental education should merely focus on increasing an individual's awareness and motivation or improve their capacity and enthusiasm for critically evaluating and changing conventional societal practices.

While zoos might mislead the public through promotional activities, they do provide an important community service by recounting the endangered species dilemma to their visitors. Providing visitors with tangible information about environmental advocacy can further enhance these civic achievements. In order to increase the likelihood that education programs will change behaviours, they will need to facilitate:

- a recognition that environmental threats are real;
- an appreciation that personal action is necessary (if people do not feel any responsibility they are not likely to take action);
- a perception of individual effectiveness (people have to feel they have the skills to take an action);
- an appraisal of the likely outcomes of action (will my actions make a difference); and
- an awareness of the consequences of their actions.[30]

In so doing, zoos would empower their visitors and demonstrate to critics and the general public that the zoo is a useful conservation organisation worthy of support.

An Enlightened Ark?

Despite the limits imposed by restrictive administrative and ideological settings, zoo teachers and other professionals continually find ways to deliver important messages about environmental issues through a range of formal and informal education programs. Some keeper talks, zoo and exhibit plans, and school materials are progressive, insofar as they specifically target some weaknesses of traditional thinking about the zoo's role in education and about the role of environmental education.

Some innovative measures in zoo planning and exhibit design are being used to deliver important environmental messages through a zoo's informal interpretive environment. Generally speaking, the major wildlife parks or sanctuaries in Australia have some interesting advantages over their metropolitan counterparts who must struggle to portray 'natural' or 'wild' settings in fairly crowded, urban sites with ageing infrastructures. They include the Territory Wildlife Park, the Alice Springs Desert Park, Healesville Sanctuary, Victoria's Open Range Zoo, Monarto Zoological Park and Currumbin Sanctuary. These properties vary in age, but nearly all of them were developed in the latter half of the twentieth century, are located in rural (and spacious) settings and specialise in native species or species from a particular region.

For example, The Territory Wildlife Park located near Darwin in the Northern Territory, features native and introduced animals of the Territory and boasts a modified open-range format on 400 hectares of tropical woodland, monsoonal rainforest and natural wetlands. Unencumbered by a historic pattern of mixed exhibit and animal collection philosophies, the staff of the relatively young Park have been able to capitalise on this unique setting and provide consistent bushland themes throughout the exhibits. The unique tropical environment of Northern Territory bushland is readily apparent at the Territory Wildlife Park. The Park's geographical context provides an unmistakable contrast to cooler, more temperate settings of most Australasian zoos, which are located in western, eastern and southern states of Australia and in New Zealand. The Park also provides an appropriate complement for tourists eager to view the Territory's wildlife, which is often not easily spotted 'out bush'. Tourists on their way to some of the Northern Territory's celebrated National Parks can incorporate a stop at the Wildlife Park with relative ease.

The Park's atmosphere is decidedly different from that of any of the metropolitan zoos. Here one has a distinct sense of being 'out bush' in the Northern Territory's unique environment, strolling through the Park's meandering paths, occasionally coming across free-ranging animals or viewing micro-habitats where animals come and go, unconstrained by enclosures. Invertebrates featured at the Park in the Arthropod exhibit represent a progressive shift away from zoos' traditional bias of exhibiting 'cute and cuddly' mammalian species. The decision to feature less winsome creatures can further conservation knowledge by acknowledging the value of diverse life forms and their importance to ecosystem functioning. In addition, the feral animal display is an another example of unique and progressive thinking in zoo exhibit philosophy. It endeavours to educate visitors about serious environmental problems plaguing the Northern Territory, the specifics of which are provided in the accompanying signs.

Talking with the Ark Crew

Many zoo education professionals believe that a human interpreter provides an important, and sometimes more effective, component to the overall learning potential of the zoo visit. It is increasingly being recognised by zoo professionals that visitors often desire more information about the animals they see and have a strong preference for interactions with animal keepers and guides. This recognition is reflected in emerging recruitment practices and industry standards in Australasian zoos. Prospective animal keeping staff are increasingly expected to have public speaking skills and to be willing to interact with the visiting public. At Healesville Sanctuary, virtually all the keeping staff take a 'public presentation' shift throughout the course of a working day. For those already employed at zoos, there is an increasing number of formal and informal training initiatives to help keepers develop their skills.

Some zoos have used keeper talks to deliver environmental messages that might facilitate cognitive and affective changes in the audience.[31] Melbourne Zoo's very popular seal show uses the theme of marine pollution to send an environmental message to its visitors. The keepers direct the performing seals to retrieve pieces of plastic and other debris that were placed in their pool before the show starts. When the seal returns to the training platform, the keepers' narrative discusses how these animals are often badly injured or killed by these items. The aim of the exercise is to encourage visitors to acknowledge the impact that irresponsible behaviour has on the environment and to encourage them to think twice before throwing garbage in the open seaway or stormwater drains.

School Programs on the Ark

Education professionals spend many hours designing support materials and programs that are engaging and enjoyable for school children visiting the zoo. A divergent range of topics is on offer, some of which specifically target critical thinking about and advocacy for environmental problems. At Taronga Zoo, education staff produce a teaching resource booklet focusing on threatened and endangered animals. The booklet discusses how teachers can use the zoo to teach their students about the very real, human-induced causes of species' decline. The booklet also offers some locally based solutions such as how to attract animals to gardens or ways of promoting the conservation of a particular animal threatened with extinction. Another teaching guide focuses on Environmental Studies. Students choose from a selection of environmental issues and are asked to consider the causes and effects of that problem, as well as positive and negative outcomes of certain responses to it.

At the Adelaide Zoo, a member of the education staff developed a role-playing exercise based on the plight of the Sumatran Tiger. After reviewing background materials on the issue, students play the roles of different stakeholders (government representatives, local villagers, business and community leaders) in a mock meeting about deciding the feasibility of purchasing a strip of land to connect two separate tiger reserves. At the conclusion of the meeting, students are meant to discuss what impact their decision might have had on the tiger population, environment and various stakeholders. The exercise uses a real example to help students develop critical thinking about the complexity of economic, social, political and environmental considerations involved in wildlife conservation.

At Healesville Sanctuary in Victoria, education staff also use a range of creative exercises in their school materials. The 'Planet Earth' program introduces primary school students to concepts of interdependence, competition and the fragility of the Earth's life systems. Students are encouraged to think about their own responsibility while ticking off 'good' and 'bad' environmental habits and reading about ways they might break their 'bad' habits. This program also facilitates critical thinking about zoos by asking students to discuss, not only how keepers have tried to make an animal's enclosure *look* like a natural habitat, but also how that enclosure is *different* from the 'real' habitat. There are similar programs for high school students. 'Here Today, Gone Tomorrow' includes sessions about the ecological changes causing some species to be threatened with extinction. Students are required to think about the problems posed by restoration programs and what implications there are for them to act on the situation. In 'Restoring the Balance', students focus on some of the moral dilemmas posed by introduced species, urbanisation, pollution, hunting and pet management. Assessment for this program is not based on students' ability to produce 'right' or 'wrong' answers, but on how well they recognise the complexity of environmental issues.

Conclusion

Encouraging more progressive environmental education in zoos might be considered unrealistic by some zoo professionals who seem to feel that these approaches are too confronting or challenging for the public and have limited commercial potential. When interviewing zoo staff, it was not uncommon to hear arguments like this:

> ninety-five per cent of zoo visitors come for recreation . . . in 50 years time they may come for conservation . . . so we don't want to unsettle them . . . [we have to] try to hold on to the exuberance of conservation staff.

Perhaps those staff that are most reluctant to shift zoo education away from traditional messages towards 'the more serious' environmental focus are more like the allegedly uninterested zoo visitors than they might care to admit. By discouraging more extensive programming in environmental education some zoo staff are, in effect, advertising their lack of appreciation for the seriousness, complexity and far-ranging nature of environmental problems, and exhibiting their low motivation for learning more about these dilemmas and for changing their own behaviours.

These trends are closely tied to and reflected in some structural features of zoos, the topic of the next two chapters.

5

Arrangements on the Ark

The development of organisational structures may seem a rather dull subject when compared with the more exciting breeding and reintroduction programmes. However, I believe that the successful structuring of organisations is fundamental to long-term efficient utilisation of [zoo] resources.

Roger Wheater[1]

When Noah was busying rescuing and restoring life on Earth, he was unfettered by concerns about a lack of space on the Ark, how to integrate and co-ordinate his efforts with other Ark-type vessels, how much he should diversify his mission or how he could best exhibit and promote his animal cargo. If Noah was unbothered by all these issues, chances are he did not have to worry much about how to order his manifest and the growing number of human passengers on the Ark, so that a smooth passage and successful mission could be guaranteed.

The international zoo community's ambition to implement a conservation role presents an enormously complex and challenging task. The modern Ark has to be kept afloat for considerably longer than 'forty days and forty nights'. Hence, considerable demands are placed on each 'vessel' in the Ark fleet. Significant human and financial resources, as well as highly flexible and responsive systems are required to manage and display wild animals, conduct inter-fleet endangered species breeding exercises and design and implement education and research programs. These requirements become even more pronounced as zoo professionals try to increase the regionalisation of their animal collection plans, expand their research profile and their participation in endangered species recovery efforts, and improve their public education programs.

These challenges are created in part by the modern Ark's 'architecture' or its structural features. Zoos carry out their activities in an array of governmental, legislative, administrative and organisational arrangements. These settings in turn dictate particular features for practices and programs on the Ark. Certain organisational arrangements in zoos influence the zoo community's attempts to formulate conservation programs which embody ecological axioms. Uncovering any systemic weaknesses in Ark arrangements for policy and administration can provide a focus for improving the zoo community's efficacy for biological conservation.

The Bureaucratic Society

In order to understand the nature of modern zoo policy, it is helpful to consider how historical and contemporary trends in Western administration pose challenges for all sorts of environmental policy and management. Society began to use bureaucratic environments to organise life during the Scientific and Industrial revolutions. In the eighteenth century, a mechanistic world view dominated people's thinking. This perspective, in effect, reduced the world (and all that it included) to a machine that was made up of interchangeable and interrelated parts. Like an industrial machine, these parts could be replaced or repaired by people as needed. Non-human nature was included in this way of thinking. Previously it was thought of as a living organism, which had a reciprocal relationship with humans. With mechanisation and the later development of industrialised capitalism, non-human nature became an object to control and dominate, a commodity to be managed in an orderly and rational fashion in the name of progress and development.

This particular thinking laid the foundations for the way Western society began to organise itself. Bureaucratic forms of organisation proliferated during industrialisation. The state, economy, and public and private institutions became increasingly formalised and structured by an administrative apparatus that was (and still is) designed for calculability and efficiency. Whereas Noah might have left his fate and that of his living manifest in the hands of a higher being, notions of progress during industrialisation seemed to hinge on the value of securing and maintaining order through perfect administrative control. It was as if mastering nature depended on mastering human nature.[2] The priority of bureaucracies was to negate the human character of administrative processes while simultaneously controlling nature for the benefit of humans.[3]

Contemporary life and notions of 'progress' are still dominated by a longing to control nature and other species. We merely need look at modern organisations to see that we hardly give official life over to fate. Rather, the metaphor of the machine is used in Western society to order life according to mechanical principles. Bureaucratic ways of organising seek to standardise administrative processes in a similar manner to the way machines standardise production by striving for accuracy, clarity, order, speed, and reliability. These goals are supposedly achieved by creating a fixed division of tasks, a hierarchical system of supervision and a reliance on a rigid system of rules and regulations.

These standardised means and ends are also indicative of a particular view of the world. Problem-solving in most bureaucracies typically focuses on compartmentalising issues into distinctive units, thereby making problems more 'manageable'. The underlying assumption of this method of organising is that 'the problem' is now under the control—not of a higher spiritual being—but of an all-knowing and all-powerful 'administrative mind'.[4] This perspective both diminishes an appreciation of broader contexts and is overly optimistic about human intellectual and administrative capabilities. Our knowledge is as imperfect as the institutions we have created for problem-solving and decision-making.

Indeed, Western society's attempts to impose a rational order on the world may have propagated some of our worst environmental problems. A host of subtle and complex organisational factors have been shown to frustrate the preservation and restoration of many endangered species. It has been well documented that the United States' *1973 Endangered Species Act* has fallen short of its stated goals to restore species and ecosystems threatened with extinction.[5] This poor performance has been caused by serious institutional problems, such as scientific and bureaucratic conservatism, resource constraints, conflicting goals within and between agencies, biases towards particular species, competing interest groups, limited professional competancies (for example, rigid leadership patterns), fragmented decision-making and poor communication procedures.[6]

There are other ways that bureaucratic settings fail us. The idea that bureaucracies would always exercise formal and careful reasoning was originally promoted as its greatest strength. At that time, the problems facing society were far simpler than the challenges of today. This form of organising is less appropriate for the complex quandaries of contemporary society, such as environmental decline. Moreover, it is highly doubtful that such rational logic could be separated from the political intricacies of life. Are these organisational forms efficiently neutral or biased by allowing for the expression of particular interests while excluding others? The insistence on maintaining order and control and discouraging controversy, in fact inhibits an open airing of conflicting values and deters input from outside interests. This form of 'organisational insecurity' also thwarts creative risk-taking by rewarding non-controversial behaviours.[7]

In this way, it is possible to see how bureaucracies have been accused of being overly rigid, conservative, resistant to change, how they perpetuate social class differences, and that those problems create inequitable and inappropriate policies. Clearly then, the term 'bureaucracy' signifies more than merely a neutral structure designed for administrative and technical efficiency. Bureaucratic arrangements are forms of *social* organisation with particular features.[8] They function as a social tool *and* method of organising and pervade nearly every aspect of human society.

Bureaucracies are a significant part of the governing apparatus of virtually every country in the world, permeating public policy-making in large organisations in the public and private sectors. In these settings, a preoccupation with getting 'organised' translates into finding a structure that moulds everything into neat parts and that centralises decision-making.

In order to achieve these highly-focused centres of control, highly political processes will be operating in bureaucracies as different groups actively utilise power resources to secure their interests. The use of power is integral to understanding the functioning of organisations. Renowned organisational theorist Gareth Morgan explains that order and direction among people with potentially diverse and conflicting interests is established and maintained by particular organisational structures. Those with power will have varying degrees of influence over: determining the specifics of organisational structure, rules, and regulations; decision processes; knowledge and information; inter- and intra-organisational boundaries; managing uncertainty; technology; interpersonal alliances and networks—the informal organisation; counter-organisations; and symbolism. We can then assume that power is expressed in myriad ways in organisations and that, ultimately, official polices will be most representative of those who posses and control the most resources.

Adherence to hierarchical structures and to rules in bureaucracies assist senior levels of administration in controlling organisational resources. Whether they are government employees, or staff of private companies or universities, senior officials play an important part in policy-making because they can mediate social and economic influences.[9] By virtue of their positions of authority, they are empowered in ways that enable them to advance their interests more readily than those with fewer power resources. For example, the presence of commercial overtones in zoo policy may be indicative of the primacy of an economic rationality that is often favoured by senior administrators (see below and Chapter 6).

Given the predominance of bureaucratic ways of knowing and organising, it would be unrealistic to assume that the international zoo community has remained untouched by such influences. The features of bureaucratic environments are manifest in many forms, such as international and regional strategies and policy-making exercises, international collection planning and exhibit design trends, and specific administrative arrangements in Australian zoos.

Bureaucracy in the Ark

A defining characteristic of the modern zoo has been *how* to select *and display* living nature. During the Scientific and Industrial revolutions, the number of zoos grew, as did a push to formalise these collections of nature and to have them reflect principles that were far more orderly and scientific than Noah's 'two of everything' philosophy. Today, the intricate administrative arrangements used to plan the selection, display and breeding of zoo animals continues to objectify nature by compartmentalising wild animals into intensively controlled environments and assigning them worth according to their capacity to serve some purpose. These exercises also reflect a certain naive optimism about how well increasing organising will serve educational and conservation goals.

One example of such ordered thinking is the World Zoo Conservation Strategy that was released in 1993. It seeks to conserve biological diversity and improve standards of care and purpose among the international fleet of Ark vessels. Yet the Strategy relies on the kind of positivist reasoning that is manifest in zoo research programs. 'Nature' is something that can be 'managed' more effectively merely by intensifying efforts to organise existing and potential members of the global zoo community. Zoo animals are referred to several times in the Strategy as 'living material', illustrating an instrumental world view which attributes worth to non-human nature on the basis of the benefits it confers to humans. It states that 'an important premise in determining the composition of the [zoo] collection is that every animal must have a function within the framework of the objectives of that zoo'.[10] It also calls for further strengthening and expansion of the global zoo network. Collectively, these statements reinforce notions of nature as an object (separate from humans) and assume that more formalisation and administrative efficiency will help zoo professionals achieve their goals. The architects of the Strategy would certainly appreciate the complexities of conserving biodiversity and of conservation planning. Nevertheless, the formal Strategy lacks a frank discussion of how complex the problem of environmental degradation is, nor does it highlight the political obstacles zoo professionals might face in their attempts to standardise zoo policy.

The bureaucratic principles underlying these broader zoo policy trends have also influenced the blueprints for modern zoos. Several organisations in the Australasian zoo community display similar 'architectural' features. Even though these arrangements are alike, they have typically been neglected in debates about the zoo's efficacy for environmental advocacy, education or conservation. This is somewhat surprising given that particular structures will order departments and employees in ways that help to reinforce the priority of certain aims and objectives.

Eleven of the more influential zoos in the Australasian region include a mix of State government instrumentalities and private organisations. The Zoological Parks Board of NSW operates Taronga and Western Plains Zoos. The Zoological Parks and Gardens Board of Victoria manages Melbourne Zoo, Healesville Sanctuary and Victoria's Open Range Zoo. The Zoological Gardens Board of Western Australia oversees Perth Zoo. These are all statutory authorities. In the Northern Territory, the Parks and Wildlife Commission manages two zoos: the Territory Wildlife Park and the Alice Springs Desert Park. The National Trust of Queensland operates the Currumbin Sanctuary and in New Zealand, the Auckland City Council operates the Auckland Zoo. In South Australia, the Adelaide Zoo is operated by a private organisation—the Royal Zoological Association of South Australia.

In Australia and New Zealand, the major zoos have compartmentalised and hierarchical organisational structures. Figure 3 summarises the administrative arrangements characterising major zoos in the Australasian region. While some zoos may differ in detail, the illustration remains largely representative of the general organising principles they use. Interestingly enough, these arrangements seem to parallel the highly-ordered, systematic patterns of their collection plans which are based largely on taxonomic classifications for animals.

Figure 1 also shows that zoo architecture typically follows the basic principles of bureaucratic structures. Personnel are arranged on the basis of a hierarchy. Like the different levels of biological diversity (see Chapter 2), the three main levels comprising the hierarchical chain of command might be likened to decks on a ship. The

Figure 1 A typical Ark structure

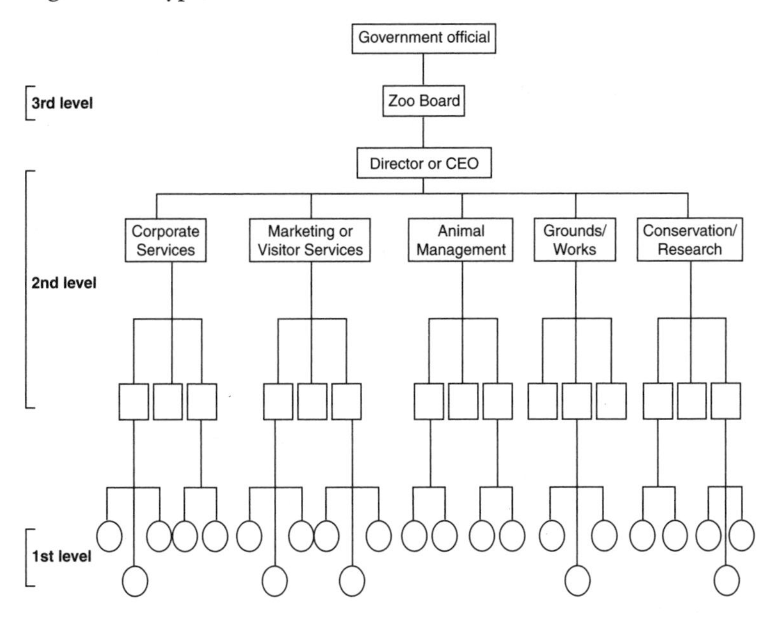

modern Ark typically has three main levels of personnel: the technical or operational level, the administrative or managerial system and the institutional level.

In most of the major Australasian zoos, the first and lowest level, is personnel who toil in the engine-room to ensure the vessel is operational: the animal keeping and horticultural or grounds staff, as well as some education officers and most clerical staff. The next and second level, is the managerial system. Depending on the size of a particular zoo, there will be several layers of management. The Zoo Director or CEO is at the top of the managerial hierarchy and oversees the total operations of their local fleet or single vessel. Underneath will be their first mates: either Directors or Senior Managers who head each of the major zoo divisions. The CEO and his/her Directors typically form the Executive Management Team. The divisions headed by the Directors or Senior Managers are typically broken down into sub-departments, the heads of which will form another level of senior or middle managers. Again, their

A koala at Currumbin Sanctuary, Gold Coast, and tigers at Western Plains Zoo, Dubbo. Charismatic fauna, be they exotic or native, are assured a place on the Ark because of their popular appeal.

classification will depend on how large or small the zoo is. Staff in the managerial system participate to greater or lesser degree in controlling the flow of information and primary decision-making processes throughout the zoo and in their respective divisions. The third level—the zoo boards—acts as a kind of intermediary between the management system and 'higher order' community interests which it is supposed to serve. Also at this level are the ministers and other government officials who liaise with zoo board members and the upper echelons of zoos' senior management strata to ensure that the political and administrative priorities of the government in power are adhered to.

Most zoos also use bureaucratic methods of organising to arrange the variety of activities they undertake. As they are seen to be part of a service industry, administrators use a product form of organisation to group activities around the provision of multiple products or services that create 'outputs'. At the bottom of the ship's hierarchy, in the cargo hold, are the Ark animals. They are the primary 'product' displayed as a service to a fee-paying public. While different departments have varied functions, ultimately these units must support zoos' commercial imperative. Do these functional divisions of labour in zoos create particular cultural and logistical obstacles for the staff who wish to fulfil the zoo's conservation aims?

Operations in modern zoos are typically organised into several separate divisions: animal management, conservation, corporate services (administration and finance), marketing or visitor services and works/grounds (Figure 1). Animal management departments are directly responsible for caring for animals and in some cases they may also manage the botanical collection as well. Staff in these departments plan which animals the zoo will hold and attend to the basic husbandry requirements of animals. In more recent decades, their activities also include managing co-operative species propagation schemes and fostering some research activities. Some of the major Australian and overseas zoos have created separate Conservation Units. These divisions operate varied conservation programs and may try to achieve greater integration and/or co-ordination of conservation and research activities across the zoo(s). Chapter 4 outlined how Education departments, which are now typically part of Marketing/Visitor Services divisions, have historically played a key role in zoo conservation by striving to facilitate an increased awareness of the need to conserve wildlife and their habitats. Education staff design and implement formal and informal education programs for schools and zoo visitors.

Other zoo departments have a less obvious connection to the conservation function. Nonetheless, these areas influence the design and implementation of conservation and education programs. Commercial matters in zoos are primarily the responsibility of Marketing or Visitor Services departments which are generally concerned with generating financial support for the zoo by increasing attendance rates, developing fundraising programs and promoting the recreational, educational and conservation functions of the zoo to the local community and to tourists. Works or Capital Development departments are responsible for managing certain components of the zoo landscape, such as co-ordinating the construction of new exhibits and maintaining existing facilities. In some zoos these departments will also include the maintenance of a zoo's horticulture collection. Corporate Services departments or Central Administration units provide an array of human resources, financial and information management services, and, where applicable, will co-ordinate the provision of State government funding.

In some of the larger zoos, the major program areas will be broken down into more specialised sections. For example, the marketing or visitor services department will have separate sections that administer public relations, media and sponsorship matters. In contrast, smaller zoos may incorporate marketing into administrative and finance units. Animal management divisions will often include several subdepartments such as animal keepers, curatorial services and veterinary care. Keeping staff are typically assigned to sections which correspond to taxonomically-related species (primate, carnivore and ungulate sections) where they develop highly specialised husbandry skills needed to care for those species.

Not all program areas of a zoo are located on the same level, nor do they receive equal accommodation. The hierarchical delineation of authority and functional division of tasks in zoos are often reinforced by different symbols of power, such as a program area's location and facilities. It is not uncommon for the different program areas to be housed separately and in different locations from each other. For example, at Perth Zoo central administration, marketing and visitor services occupy separate buildings, but both are placed at the front of the Zoo. The animal management department is located at the opposite end of the property, with several of its subdepartments housed in different areas of the Zoo. The site for Taronga Zoo in Sydney is essentially a hillside. The authority of central administration is reflected in part by its location in a separate building at the top of that hill. Animal management units are scattered throughout the property, and the research and conservation units are in separate buildings at the bottom of the hill.

Decision-making in and around the community of zoo professionals reflects the fact that the preferences of some will eventually be adopted over others. Power involves relationships between political actors, be they individuals, groups, or other aggregates. These exchanges do not eliminate the possibility that such inequalities may be resisted by the less powerful, nor does it deny that the more powerful may feel some regret using their advantageous positions. Nonetheless, organisational arrangements are critical to the use and determinants of power:

> [Power is] a structural phenomenon created by the division of labour and departmentalisation that characterises the specific organisation or set of organisations . . . although individual characteristics will affect the exercise of structurally-determined power.[11]

To gain further insight into how bureaucratic structures favour particular values and affect contemporary zoo policy, it is helpful to consider the shift in zoo administrative trends in the last thirty years.

The change in the size of the global zoo community and of individual zoos has had significant impacts on the overall zoo conservation mission. During the second half of this century, there have been intense pressures on zoos to respond to shifting societal values and ecological conditions. It has been a time of formidable development and growth for the international zoo community. The proliferation of zoos and zoo infrastructure has coincided with zoo professionals' growing concern about how they ought to incorporate environmental values into their institution's râison d'être. The following discussion uses the example of three major zoological park boards to trace how these trends have been manifest in zoos. As these organisations have grown, they have increasingly formalised new policy goals, specialised divisions of labour, and expanded their administrative, commercial and senior management areas. It is almost as if the Ark has been filling up with more and more people. 'Noah' now has less time to deal directly with his animal passengers, as he is too busy trying to decide how to allocate resources and tasks across his ever-expanding fleet and crew.

The 1970s

When the *Zoological Gardens Act 1972 (WA)* was passed in Western Australia, the Perth Zoo's links with the State's National Parks Board were broken and a formal Zoo Board was created. In its annual report for 1970, the Zoo Board first reported that its primary purpose was to maintain and display animals, particularly Western Australian fauna, for educational purposes, to undertake research, and to breed rare species. A veterinarian, Tom Spence, was appointed as the Zoo's first professional director. He sought government support for a major redevelopment scheme and introduced a quota system for determining which species and individuals would be held and displayed. There was no 'senior management structure' *per se*, but an overseer, head gardener and maintenance foreman reported directly to the Director. Of the total staff of forty-three, twenty-five were grounds and maintenance workers and the remainder were animal keepers.

By 1975, the Perth Zoo staff had grown to a total of fifty-seven with increases primarily in animal keepers and grounds personnel. Its relatively simple organisational structure had not changed significantly. However, education now featured prominently in zoo policy when the Director General appointed two teachers to coordinate activities at the Zoo site. The year's capital works improvements focused on provision of a new administration building. The Zoo's Board continued to comprise a mix of prominent citizens from senior levels of the State's public service (in the areas of agricultural and wildlife), university (zoology) and business communities.

In New South Wales at the beginning of the 1970s, Taronga Zoo was still overseen by the Taronga Zoological Park Trust. Similar to Perth's Zoo Board, Trust members included an array of distinguished citizens specialising in law, business administration and the agricultural and biological sciences. The Director at this time was a prominent scientist with a background in zoology. In these early days, the Taronga Trust reported more specifically on its organisational structure and size than Perth's Zoo Board did. Taronga Zoo was significantly larger and had a more complex organisational structure than its Western Australian counterpart. There were 150 staff assigned to three main divisions: the Director's section and the Administrative and Exhibits divisions. There were more senior staff than the other zoos, including an Assistant Director with expertise in administration, a veterinarian, several curators, a senior education officer, a press officer, works and operations supervisors, an office manager and a catering manager. There were also already several professional levels within the animal keeping department, with Curators advising on planning animal collection, a superintendent of exhibits, head keepers, department keepers and two further grades of keepers. A similar structure was emerging among horticultural staff.

In the mid-1970s, the Taronga Trust had already been dissolved and replaced by the Zoological Parks Board of NSW. This statutory body had now grown to eleven members, several of whom had previously sat as Trustees for the Zoological Trust. The Board's responsibilities now included developing and managing a second property, the Western Plains Zoo in Dubbo. This zoo had twenty-five staff, the most senior was its Curator. Meanwhile, the staff of Taronga Zoo had increased by ten. There were slight but significant shifts in the Zoo's organisational arrangements. At the beginning of the decade, Education and Public Relations had been combined into one department with seven staff. These two areas were now separate departments with six and five staff, respectively.

Like Western Australia and New South Wales, the State of Victoria took steps to legitimise the role of its zoos when the

Zoological Parks and Gardens Act 1973 (Victoria) was passed to establish its Zoo Board as a statutory authority. The Board would oversee the management of three properties: Melbourne Zoo, the Sir Colin MacKenzie Zoological Park (SCMZP) and the Werribee Zoological Park. The Act was amended by 1975 and specified the Board's general powers and functions. The three zoos would be maintained in order to exhibit and scientifically study 'zoological specimens', instruct and entertain the public, and protect and manage of wildlife and their habitats.

The 1980s

By the 1980s the three major zoos were moving towards becoming even more bureaucratised. Levels of formality were rising, as the zoos began to engage in more lengthy and detailed reporting to government regarding how their activities were helping them to achieve their aims and objectives. By the end of the decade private sector principles were infiltrating the management of the public sector, and the statutory zoos were struggling to maintain and develop their conservation role while conforming to these new administrative trends.

In the mid-1980s the size of the Zoological Board of Western Australia had not changed significantly in size. However, with the appointment of the new Director, a former curator at Taronga Zoo, the Zoo engaged in further articulation of its goals and objectives. A formal policy statement in the Zoo's annual report now listed the 'four pillars'—conservation, education, research and recreation—as its primary aims. The Zoo was to achieve those aims by focusing on Australian fauna in planning the collection, using naturalistic settings for animal displays, conducting breeding programs for rare species, and implementing education programs to foster positive attitudes to nature. These objectives had been used in the zoo community for some time, but now there were some administrative objectives that were supposed to help the zoo deliver its mission. These included achieving high performance standards, promoting the Zoo's activities, developing the Zoo as a local and State asset, and regularly reviewing Zoo policies and achievements.

The Perth Zoo's organisational structure and staffing reflected the need to care for a growing animal collection and to fulfil its education imperatives, but it also now reflected some growing administrative burdens. Overall, the Zoo had grown to seventy-two staff. The animal keeping section had increased from seventeen to twenty-five keepers and had three professional levels (head keepers, Grade 1 and Grade 2 keepers). The Zoo's Curator was considered a senior staff member and reported to the Director. There was now an Animal Health department, with two full-time support staff and a Veterinarian who also acted as one of two Assistant Directors. The Zoo also had a separate Education staff, although its five staff were on part-time duties. The Zoo's second Assistant Director was overseeing a range of administrative matters. A Visitor Services area had been established, but it was essentially filled with part-time staff.

In NSW, there was another shift towards more formalised policies and procedures. One indicator of this trend was the changing format of the Zoo Board's annual reports. The 1981 and 1982 versions had a distinctly comic style, featuring large letters and photographs and cartoon characters in its graphics. By 1984, the photos of varied zoo activities remained. But a more prim and concise account of the Zoo Board's powers, as set out by the Act of 1973, and its primary and secondary goals, had replaced the humour. By the middle of the decade, the annual reports included a full list of the staff at the Zoo Board's two properties (Taronga and Western Plains Zoos). This format remained largely unchanged for the rest of the decade.

Taronga and Western Plains Zoos had also grown considerably. At Taronga Zoo there were 229 staff (including part-time and casual staff). While all Taronga Zoo's thirteen departments had grown, some of those increases and the creation of new departments signalled the rising importance of administrative, promotional and commercial activities. The Public Relations, Admissions and General Office/Accounts departments had all more than doubled in size. The Trading department had increased to ten times its size in the mid-1970s. Taronga Zoo now also had a separate Sponsorship department. At Western Plains Zoo, the total staff had risen to thirty-nine. New departments included Souvenir/Catering (eight staff) and Admissions (three staff).

Alongside the increase in staff came a further delineation in the layers of authority. What had been the 'Director's Division' was now the central administration area and included the Director of both Taronga and Western Plains Zoo, a Deputy Director (who supervised operations at Western Plains Zoo), an Assistant Director, a Public Affairs Director, a Personnel Officer, as well as several clerical staff. Within the Animal Department, several more layers of management had been designated, including a Veterinarian, three Curators and a chief of operations. Under these staff were supervisors of mammals, birds and the aquarium, as well as department keepers and section heads. Even Western Plains Zoo, which was considerably smaller than its city counterpart, now had three senior staff in its Administrative unit.

At the beginning of the decade in Victoria, the Zoological Board's composition still reflected a range of expertise which had been established by the original Act and subsequent amendments. The Chair had strong zoo and conservation qualifications. Several members possessed veterinary and other scientific skills, given their work in academia and museums. There were also representatives from local and state government and the professions of architecture and engineering.

Like its counterparts in WA and NSW, the Victorian zoos reflected growth in their overall size, further delineation of authority structures and the beginnings of centralised decision-making. The increase in administration was in part reflected by the growth in the number of Board Committees that had doubled since the beginning of the previous decade. The Executive, Scientific, Education and Publications committees were still in place. What had been the Works/Finance Committee now constituted two separate committees, with 'works' changing to 'development'. Other additions included committees for veterinary matters, promotions and operations of Werribee and Sir Colin Mackenzie Zoos. The senior and administrative staff listed in the Board's annual report for 1986 had risen to fifty-five. The Board now had a CEO and Directors of each of the three properties, Melbourne Zoo, SCMZP, and Werribee Zoological Park, who reported to the CEO. In addition, a 'Group Zoologist' and Chief Education Officer answered directly to the CEO and were considered part of the Central Administration.

The divisions for the properties were somewhat simpler than in NSW. Both Melbourne Zoo and SCMZP each had Administrative, Animal, Horticulture and Works Divisions headed by a senior or middle manager. Each property also now had a Business Manager who conceivably was to oversee a range of financial and commercial matters. The Education Department staff provided services for Melbourne Zoo and SCMZP. The Zoo Board contracted services from consultants who had expertise in retail, horticulture and graphics. Given Werribee Zoo's small size, there were no separate departments *per se*, but there were three supervisory staff including a Manager, Assistant Manager and senior keeper. Like Taronga and Western Plains Zoo, the Animal Division had now designated separate Curators for birds and animals, as well as head keepers for different animal sections, such as carnivores, primates, native mammals, etc.

The appointment of a new CEO in 1986 was a significant departure from traditional appointments of people with some sort of scientific knowledge or zoo management experience. Mr. Bob Carr, not to be confused with the Premier of NSW, had extensive business qualifications. He had worked as a senior financial analyst, a manager for corporate planning and development, and a general manager for several large corporations such as BHP, Elders IXL Ltd and Kemtron Ltd. Mr. Carr's position foreshadowed what would become a trend: highlighting the importance of managerial concerns in organising zoo policy by hiring CEOs who lacked academic or professional qualifications in science or zoological management, but who had strong backgrounds in business management. Like the NSW Zoo Board, the Victorian Board began to use a more stylised graphic design and layout for its annual reports and included increasingly formalised and detailed accounts of its functions and activities. By the end of the decade, the inclusion of organisational flow charts in the annual reports signalled an increased focus on administrative strategies and planning.

The 1990s

In the 1990s the 'big three' Australian zoos continued to grow and rely on managerial mechanisms to fulfil their official (and unofficial) aims and objectives. High ranking officials' use of legislative tools, significant organisational restructures, the appointment of professionals with business management expertise to senior zoo positions, and increasingly stylised and detailed annual reports and other forms of promotion have all provided some indication of the strength of this trend. These strategies have been more supportive of certain values than others.

During the early 1990s the Perth Zoo had nearly doubled its total staff numbers since 1985 and had streamlined its organisational arrangements. There were now five main program areas: Collections, Marketing, Education, Corporate Services and Park Facilities. Managers of each of those departments and the Director formed the Zoo's executive management team. The Director of ten years, John De Jose, remained highly focused on consolidating and elevating the conservation profile of his organisation and the zoo community.

Paralleling a concern with fulfilling a conservation mission, was a growing interest in administrative efficiency and visible success indicators. The Zoo Board was now chaired by a leader in the business community. He signalled his economic concerns and faith in organisational planning instruments as means to realise Zoo goals:

> [The Board was] faced with the challenging task of consolidating Perth Zoo's position as a conservation force and ensuring the Zoo's financial survival as an operating business. Review of the Master Plan and Business Plan commenced during the year and this will establish a range of achievable objectives.[12]

The Board's annual reports, now highly stylised with a corporate logo designed at the beginning of the decade, included a new mission statement and detailed accounts how each program area was performing to the highest standards and realising a host of objectives associated with the Zoo's overall mission.

By the following year, however, the appointment of a CEO and subsequent organisational restructure intimated otherwise. A high-ranking public servant from the Tourism sector was now running the Zoo and the strategies she employed suggested a shift away from the values of the previous Director. The Zoo had become more top-heavy and centralised. The executive management team had grown to six Director positions: Research, Planning and Development, Visitor Services, Conservation, and Strategic Policy/ Management. The Directors' positions now took up—at a higher professional status—the level of hierarchy previously occupied by the senior managers. This strategy created another layer of management by pushing senior managers down into what was now, in effect, two levels of middle management.

Some imperatives were better served by this reorganisation than others. We have seen earlier in Chapter 4 that since the Education program had been subsumed by Visitor Services, it no longer had direct representation in executive management decision-making fora. The Education Manager was one of four managers in the Visitor Services program area and Education staff had been cut from 7.1 in 1994 to 4.5 in 1998. Similarly, the Collection Program and Park Facilities programs had dropped from eighty-one total staff to sixty-seven in the new Conservation program area. Conversely, the Marketing program had been staffed by 8.2 positions in 1994, and was now essentially the Visitor Services program and had eleven staff, excluding the Education staff.

A more specific analysis of zoo administrative trends in NSW during the past decade has been limited by the Zoo Board's executive management refusal to allow me to interview its zoo staff or to provide me with access to recent and specific information about financial and human resource management practices used in its two zoos. They defended their decision on the basis that there were no apparent benefits in supporting this type of organisational research and that they believed the research was 'divisive'. This kind of response supports the plethora of research about how bureaucratic organisations seek to discourage 'outside' scrutiny and the possibility of controversy. It is still possible to detect general trends in NSW that are similar to those in Western Australia and Victoria. I draw this conclusion from the evidence that the NSW Zoo Board's defensive behaviour provides, from earlier interviews with Taronga and WPZ staff that I *was* granted permission to conduct, and from content analysis of some publicly-available materials such as the NSW Zoo Board's most recent annual reports.

By the 1990s bureaucratic styles of administration and management of Taronga and Western Plains Zoos had been in place for some time. The use of intricate reporting systems were apparent given the length and density of the 1992 and 1997 annual reports which are 144 and 120 pages, respectively. The use of performance measures, indicators and targets for the zoo 'industry' had been first developed and promoted heavily by the Zoo Board's CEO and staff from Taronga Zoo.

Trends in staff numbers from different program areas suggest shifting priorities. The priorities for conservation, public education and environmental programs can be partly demonstrated by the split of the Conservation Research program of 1992 into two separate programs, the Conservation Research and Environmental Management. Each of these programs is headed by a member of the senior executive. However, there has not been significant growth in either area between 1995 and 1997, and their survival has depended largely on obtaining funding from outside the Zoo Board's revenue stream. There are other trends worth considering. Early in the decade, the Life Sciences program area had the most staff (159), followed by Corporate Services (118) and Commercial/Visitor Services (116). By 1997 Corporate Services had grown to the largest program area with 195 staff, and the newly-named Marketing division now had 140 staff. Meanwhile, total staff numbers in the Life Sciences area have been declining since 1995, and have been below 1992 levels.

By the late 1980s and early 1990s there were signs that administration of the Victorian zoos was also being increasingly challenged by competing priorities. The Chairperson of the Zoo Board signalled what would be a growing concern with efficient administration:

> resources have tended to diminish under the burden of [prevailing] economic conditions . . . the long term future of the Board's properties is a cause of considerable concern given its continuing need to rely on Government funding for about forty percent of its recurrent expenditure.[13]

At the request of the State government, the Board had prepared a three-year Business Plan that aimed to provide specific parameters for its financial planning.

Despite these concerns for operating costs, the organisation was becoming top-heavy and there was greater representation of administrative and commercial issues than there had been the previous decade. In 1992, the executive management team of the ZBV now had seven positions, in addition to the CEO. Each of the Board's properties had Directors. Melbourne Zoo had two Assistant Directors, one who also acted as Marketing Manager for the Zoo Board and the other who also served as Human Resources Manager for the Board. The Principal Education Officer retained representation in this decision-making strata, and there was also a Manager of Information Services for the Board's three zoos.

By 1993 there was further evidence of increased centralisation and shifting priorities. The CEO's responsibilities now included acting as Director for Melbourne Zoo, where the Managers/ Assistant Directors reported directly to him. The Principal Education Officer was still represented at this senior level of decision-making and had been elevated to Assistant Director. The growing need to demonstrate a priority for conservation was addressed with the appointment of an Associate Director of Conservation & Research.

By the middle of the decade, the concerns for managerial accuracy and business acumen were formalised by legislative decree and subsequent organisational restructuring. With the passing of the *Zoological Parks and Gardens Act 1995* (VIC), a new Board and Chairman were appointed. The Board no longer included representatives from the scientific (veterinary/zoology), academic and museum communities, nor was their representation from local government. The Board had been reduced from eleven to nine positions and was almost entirely comprised of professionals with business management qualifications.

The organisational restructure reflected even greater centralisation, with some departments losing representation in this critically important level of decision-making, while the influence of other areas had increased. The CEO's position no longer included directorship of Melbourne Zoo. The executive management strata included one less position and was now comprised of six Directors. Staff in the Assistant Director positions had been elevated to

Directors of their own program areas, Business Services/Planning and Marketing/Visitor Services. A Director of Conservation/ Research was appointed, increasing the representation of the zoo's important function in the most senior levels of the organisation. This effort to promote conservation stands in stark contrast to the Zoo Board's support for Education. Chapter 4 documents how the position of Assistant Director of Education had been eliminated, resulting in the loss of direct representation for education imperatives in executive management decision-making fora. The Education Department was now lead by a middle manager reporting to the Director of Marketing/Visitor Services.

This reorganisation, in effect, was similar to the restructure of Perth Zoo. It created another layer of management by decreasing the authority of some senior management positions and creating almost another layer of middle management. The Directors of each of the Board's properties were now lower in seniority than the other Directors. In addition, their authority has been usurped by the fact that their respective departments such as Marketing or Finance report sideways and upwards to the Directors of Marketing/VS and Business/Planning, respectively. These arrangements have resulted in unclear reporting relationships and have slowed the flow of communication. While some staff have taken upon themselves to report both to their Zoo Director as well as to the Directors of Business Services/Planning or Marketing/Visitor Services, their ability to do so has been hampered by high workloads and time constraints.

The Social Ark

The cultural aspects of an organisation lend further insight into how these bureaucratic features might frustrate zoos' achievements in conservation and education. Many organisational theorists have depicted organisations as social systems, which are created for the purpose of achieving goals. Collections of individuals communicate with each other in a setting where there are collective aims and a functional division of labour put in place to achieve those goals.

One way of thinking about how zoos operate is to picture an iceberg (see Figure 2). The formal (and visible) aspects of a zoo form the tip of the iceberg: its goals, technology, structure, resources, staff, as well as its animal displays. Hidden from view, below the water line, lie the behavioural (or informal) elements of the zoo. The fact that it is harder to see the informal aspects of an organisation does not detract from how central they are to the effective functioning of that organisation. In some ways, the informal aspects of a zoo, or its organisational culture, lend greater insight into the nature of the more immediately visible characteristics. Superimposed on the official blueprint of any organisation and its environment is a more subtle network of human groups, with their

Figure 2 The zoo as an organisational iceberg

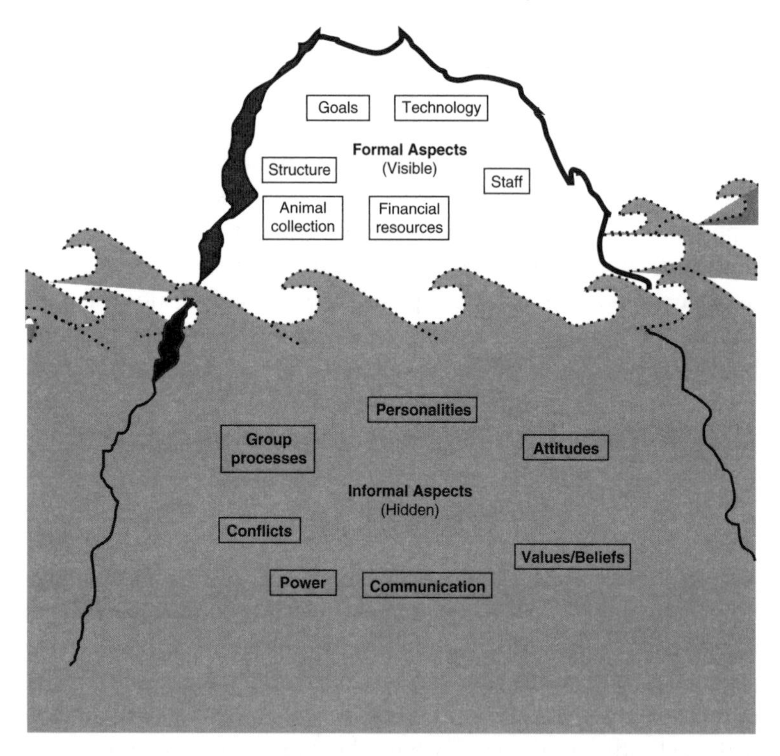

SOURCE: Adapted from D. Hellreigel and J. W. Slocum, *Organisational Behaviour: Contingency Views*, West Publishers, St. Paul, Minnesota, 1976; J. Jackson and C. Morgan, *Organisational Theory: A Macro-Perspective*, Prentice Hall, Inc., Englewood Cliffs, NJ, 1982.

loyalties, prejudices, antipathies and codes of behaviour.[14] An organisation's culture are those philosophies, ideologies, values, assumptions, attitudes and norms that people use to attribute meaning to events and to make sense out of their worlds.[15] An organisation's culture is very relevant to the design and successful implementation of policies and strategies. For example, zoo staffs' ideas about appropriate conservation activities for their institutions significantly influence the design and implementation of its programs.

It is valuable and necessary to differentiate work in organisations and have positions of authority to maintain some degree of consistency and control. Yet, in the zoos I studied, there are signs that their transformation into progressive conservation-based institutions is constrained by the more common afflictions that result from functional specialisation in bureaucracies. Those include poor interdepartmental communication, compartmentalised understandings among staff about the functioning of their organisation, different departments pursuing opposing goals, narrowly and rigidly-defined jobs which discourage staff initiative and flexibility, and competition among and within departments for organisational resources and recognition for achieving goals.[16]

From ministers to senior public servants, and from CEOs to the animal keepers, zoos' chains of command are made up of a diverse range of people including professional politicians and bureaucrats, wildlife professionals, education specialists, marketing and business experts and veterinarians, to name a few. All of these people have both common and disparate perspectives on the roles of zoos. The commonly held views hold zoos together by helping to achieve policies that are consistent with zoos' formal goals. That is, many are largely supportive of zoos and believe the World Zoo Conservation Strategy embodies appropriate goals for the zoo community. There are also points of difference. These contests contribute to practices that contradict zoo aims, or that somehow misrepresent actual practice. Some individuals occupy formal and informal positions where they are able to make their views known, others have little access to decision-making fora. Hence, zoo policy is likely to be most representative of the powerful and influential zoo staff, board members or other government representatives.

These dynamics can fragment zoo policy and alienate employees in different departments. The organisational structure of the zoos and the resulting specialisation of functions of the animal management and other departments, facilitates particular cultural groupings or subcultures within and across zoos. Staff frequently refer to themselves according to their respective subdivisions or subsections and socialise accordingly. These subcultures often coincide with particular views about suitable conservation roles and practices for zoos. While zoo subcultures are not inherently problematic, in the context of rigid organisational structures and strict hierarchies of authority, they can contribute to inferior organisation-wide communication and disheartened morale. Currently, there are many zoo personnel who are worried about the effects bureaucratised structures and commercialised policies might be having on their organisations.

In 1994–95, I interviewed a total of 126 staff from nine Australian zoos. In eighty of those interviews, zoo staff consistently made reference to bureaucratic structures and ideological divisions that they believed compromised communication, as well as the conservation, research and education goals of their organisations. One middle manager observed how 'the grand plans of management . . . filter down and become logistical nightmares for lower levels of staff [because] not enough time is spent thinking about how these "great" ideas are going to actually work'. Noting how there was 'more bureaucracy here', relative to other zoos, one keeper spoke about a 'facade of an open-door policy' by senior management, and he questioned how sincere this 'willingness to listen' really was. 'The bureaucracy and red tape' also frustrated another keeper. She believed that 'there's far too much of it . . . [so] things move too slowly'. A middle manager saw the zoo as 'a very political institution', where the central mission was 'mostly about pleasing ministers', and there was 'little curatorial input or receiving of information' relating to major decisions. On a similar note, another middle manager believed that curators ought to have a greater say in zoo policy-making given that 'the top is run by administrators who are not animal people'.

Conflicting values across organisational levels and divisions arose continually in the interviews. One keeper who saw himself as a 'dedicated conservationist' believed that 'the zoo is run by people who are more like bureaucrats, politicians . . . or those with [strictly] financial interests rather than those with any real biological knowledge'. Another keeper felt an appropriate solution to such a dilemma would be to 'change the attitudes of the people high up . . . those who see the zoo mostly as a tourist venture'. One middle manager grappling with the implications of her recent promotion said, 'I'm a manager now, I have to stop thinking like a keeper'. An awareness of ideological contests was not limited to the lower and middle organisational levels of the zoos. One senior manager was sympathetic to middle managers' concerns and saw them acting as 'the mechanism [of] . . . integrity against too much commercialisation' in the zoo. Another senior manager felt that 'someone has to be business-oriented, but [that] should be tempered by someone in an influential position with [knowledge of] moral issues relating to the welfare of animals'. In Chapter 4, we saw how zoo staff disagree about the influence of zoo promotions and commercial programs and their potential to mislead the public. Similar issues continue to worry zoo professionals.

In 1999, 57 per cent of zoo staff at one of Australia's largest zoos reported in their interviews with me that the most significant problems preventing their zoo from achieving its mission were organisational dysfunction and a peculiar culture of decision-making (see Table 5). These staff listed several reasons for these problems persisting. They felt that certain members of the executive management strata were unco-operative and insisted on dominating policy-making processes. Not surprisingly, they suggested that these attitudes needed to be reformed and be replaced by a more open and tolerant decision-making system. However, these staff realised that before such reforms could be implemented, many people's concerns about losing power and control would need to be overcome.

Another related problem listed by zoo staff was the particular culture of decision-making in their organisation. Zoo staff felt that dominating commercial and business interests and a feigned

Table 5 Zoo failure, as assessed by staff

Issues	Causes	Solutions	Obstacles
Shortage of conservation and animal specialist skills	• political, cultural and physical isolation of zoo • historic recruitment practices • poor working conditions		• insistence that all activities have revenue-raising capacity • lack of funds
Lack of regional planning	• political, cultural and physical isolation of zoo • organisational structure favours business interests	• re-evaluate organisational structure and increase representation of non-business interests	• corporate plan locks zoo in overspending cycle • lack of funds to increase conservation staff
Problems with organisational structure	• senior managers and Board unwilling to recognise problems of excessive centralisation and benefits of redressing problems • predominance of business values • reflection of broader societal trends	• open acknowledgement of problems, review existing structures and share findings with all staff • encourage greater vision and risk taking • continue to deliver high quality species management programs	• people fear a loss of power and control • insufficient appreciation of other standpoints
Dysfunctional culture of decision-making	• dominance of commercial/business interests and its associated performance criteria • an interest in status of conservation without full commitment to it • decisions reflect societal obsession with consumerism/materialism	• reduce size of senior management strata • increase representation of non-business interests • increase organisation-wide awareness of broader environmental problems • appoint a more dynamic, informed Board • acknowledge problems, encourage creativity and vision • implement intra-zoo staff exchanges • increase conservation programs	• corporate plan reflects government's commercial priorities • rigidity of individual or specialist agendas • a repressive organisational structure • morale dampened by continual reference to the 'bottom line' and cuts to operational staff

interest in conservation were the primary cause of bad decision-making. They offered solutions such as creating more open and candid participation and more equitable representation in key decision-making fora, as well as facilitating greater inter-departmental understanding in the zoo. Zoo staff listed inflexible regimes and poor staff morale as factors obstructing these efforts at problem-solving.

These perspectives about organisational dysfunction in zoos offers a stark contrast to what is often reported in zoos' media coverage and in other forms of promotional activities, such as annual reports and other official policy and strategy documents. These documents are often slick and fashionable, and their stories convey a very particular image. While they may make some passing, general reference to 'challenging economic climates' or 'financial problems', the overall tone to these documents tends to be that each zoo is achieving all that it set out to achieve. At what cost does the push to portray success come? It may come at the same cost as when the zoo compromises its ability to promote ecological values and environmental education by relying on anthropomorphic interpretations of zoo animals or by promoting consumerism. Either way, it seems that what *is* forfeited are more open discussions about how problem-solving inside zoos is (or might be) conducted, because any signs of dissenting opinion, less-than-adequate performance or organisational unrest are kept out of sight.

Challenging Conventional Architecture

There is a certain contrariety in zoo policy. Ordered efficiency and success on the one hand are represented in zoos' contemporary administrative arrangements and policy documents. On the other hand, the views of many zoo staff suggest some poorly functioning organisational and policy-making processes. Is it merely a case of unruly Ark crew-members pushing for mutiny, or are there some substantive problems with zoos' organisational and policy-making processes? Perhaps zoos' ability to implement ecologically-inspired conservation policies has been more a function of the persistence of exceptionally dedicated and motivated individuals who are working to secure those ideals *in spite of* the structures they work within, rather than because of those structures. The cartoon displayed in Figure 3 has been circulating amongst the Australian zoo community for a number of years, and some zoo staff believe it accurately portrays not merely the exhibit design and construction

Figure 3 The consequences of functional specialisation in zoos

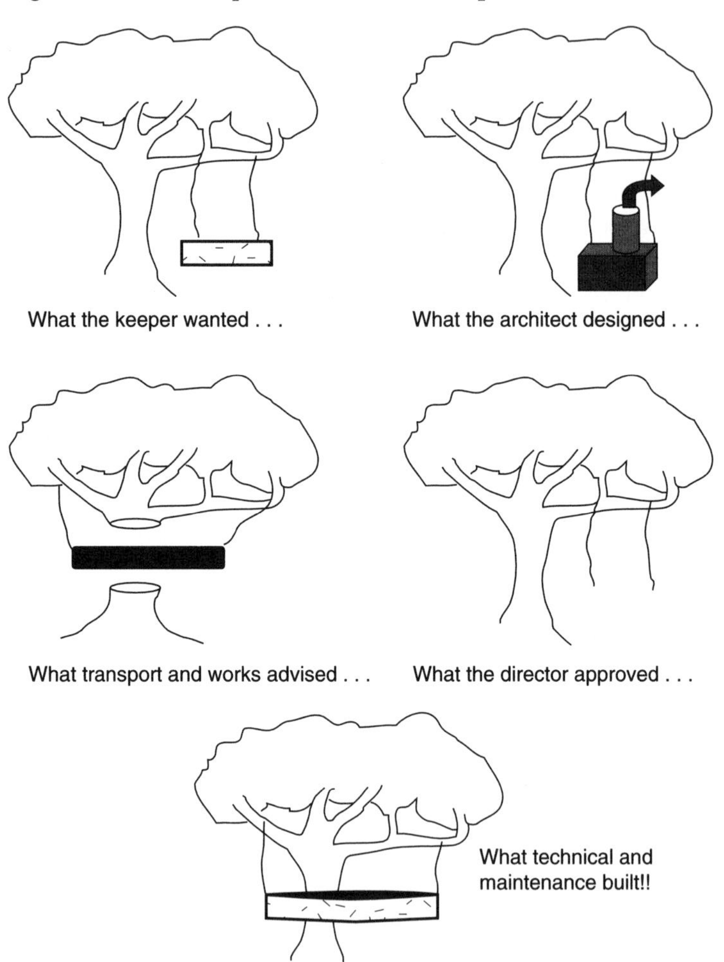

SOURCE: Anonymous

process, but the overall zoo policy process. The cartoon indicates the considerable humour and optimism many zoo staff use in the face of what they feel to be frustrating and unwieldy bureaucratic policies.

In addition to such whimsical outlooks, zoo professionals have offered constructive suggestions for addressing some of the problems resulting from bureaucratic environments. They have promoted disbanding traditional divisions between zoo departments and fully integrating their operations.[17] At an Adelaide Zoo staff workshop held in 1991, participants felt that there was a greater need for trust, consultation and sharing of decision-making between operational staff and management. These zoo staff believed in emphasising congenial and communicative relations among all staff, in addition to developing needed resources. Certainly, the smaller organisations, such as Adelaide Zoo, appear to be able to engender greater intra-organisational communication than their larger counterparts. In addition, senior managers or directors who are supportive of the need for organisational reform, will facilitate more effective processes, more readily. Table 3 shows a plethora of ideas for engendering better intra-zoo relations and performance.

I have suggested that calls to formalise and order zoo policy through ever-increasing levels of co-ordination in species management and other zoos plans may suffer from being overly-bureaucratic. The aims of programs such as the Australasian Species Management Plan and others like it (see Chapter 2), and the process of their implementation also offer some promise for better inter- and intra-organisational communications. When combined with problem-solving exercises like the ARAZPA workshop (see Chapter 2), these schemes can challenge the rigid bureaucratic structures and philosophies that might otherwise confound further integration of intra- and inter-zoo programs. Regionalising the animal collections of Australian zoos may be highly ambitious, but certainly provides a mandate for achieving excellent intra- and inter-organisational communication and a high degree of flexibility.

Other challenges to rigid thinking include the shift to contemporary exhibit design philosophies (see Chapter 4). These programs are targetted at shifting the layout of animal exhibits away from rigid taxonomic grouping of animals. Instead, habitat or ecosystems themes are used to revamp existing exhibit precincts or to develop new exhibit areas. For example, at Taronga Zoo the keepers in the Bird division used their initiative to redevelop an older area of aviaries. The original precinct was comprised of twenty-six single aviaries with a selection of different species, little or no planting inside them and no representation of biodiversity or ecological themes. The new aviaries were rebuilt at a very low cost into six larger structures with mixed species and heavily planted with native vegetation. The desired effect was to recreate a sense of continuity and connectivity among the major habitat types found across Australia (seashore, desert, grasslands, woodlands and wet sclerophyll forest and rainforest).

Arrangements for co-ordinating the Ark's fleet of vessels and the construction of individual vessels will continue to influence how close modern zoo professionals can come to realising Noah's biblical mission of restoring the Earth's ecological balance. Currently, bureaucratic structures impose a particular order and ideology on zoos (and other conservation organisations) that reflect and reinforce a highly rationalised and industrialised society. Such a restrictive framework is not necessarily equipped to face the unique and intricate challenges presented by environmental problems and it tends to devalue all but the most conventional alternatives.

Proponents of such new-fashioned options like those mentioned above, must work quite hard to see their ideas legitimised in formal policies. Certainly, zoos can benefit from implementing greater efficiencies, but what this usually means is that zoos should behave more like commercial enterprises. If we view the zoo this way, as an institution engaged in the recreation 'business', then we might consider that a zoo meets its obligations primarily by buying and selling goods. Such an imperative begs the question, what is the proper business of the Ark? If zoos are to achieve their full measure of conserving biodiversity, conducting conservation-based research and delivering environmental education, how might these tasks *differ* from those in a commercial enterprise?

6

The Corporate Ark

If only she could create a form of mystery tour, she thought, so that the zoo itself contained elements of excitement over and above the animals it housed. Maybe introduce some sort of competition. The art was to get people to spend as much once they were through the gate as they had spent getting in . . . she took him by the arm and guided his attention towards the model of the zoo spread out across the floor. 'Vince, the next thing after virtual reality is going to be reality. This could be the first of a chain of ecological zoos that people can rely on for exactly the same natural experience wherever they are—a kind of Zoos 'R' Us'.

Iain Johnstone[1]

Noah was probably more fortunate than 'Willa' and 'Vince', characters from the movie *Fierce Creatures*. He did not have to grapple with privatising and commercialising the Ark in order to find reliable sources of revenue, or with deciding how to allocate scarce resources among his passengers and crew. The mission of Noah's Ark was serving the ultimate public good—divine intervention took care of any practical or economic concerns. In addition, the simple structure of the Ark suggests some very obvious goals. Noah had the advantage of knowing what he had to do and how long it would take. He was, in effect, able to start from scratch, building an Ark that would help him achieve his aim.

Today's Noahs—the architects and administrators of the modern Ark—have been less fortunate than their biblical counterpart. There are intricate structures and conflicting goals on the modern Ark. Unlike Noah, modern zoo professionals have not been able to start with a clean slate. They have continually modified their goals over the course of zoo history, trying to agree on the Ark's course and final destination. In addition, modern Noahs have to decide how best to update their vessels' ageing infrastructure and how to reorganise staff and resources. As the number of zoos has increased and individual zoos have grown in size as well, the modern Ark's structure has become increasingly complex: goals continue to shift in response to changes in the broader social, political, economic and ecological landscape.

The favoured architectural design for organising people and resources on the modern Ark has been a bureaucratic structure. This structure is supported and reinforced in part by particular dominant management ideologies. These concepts in turn derive from broader, conventional economic theories which prescribe how society should be governed, and how resources ought to be used and distributed among institutions, groups and individuals.

To Market, To Market!

Economic and political decision-making throughout much of modern society in the last three decades has been dominated by what has been termed 'economic rationalism'. Economic rationalism has its origins in the older principles of 'laissez faire' or 'neo-classical economics', and has also been referred to more broadly as 'neo-liberalism'. It asserts that a prosperous economy depends on efficiency, and the greatest efficiency occurs when open competition in a free market determines outcomes. Because individuals are the best judges of their own welfare, markets and prices are thought to be the most reliable means for attributing value to things, and delivering better outcomes than governments which are inherently inefficient.

Almost all countries are now dominated by these market imperatives. The goal is to enable national economies to grow and compete in a global market; and market institutions, methods and motivations are seen as integral means to this end.[2] Reorganising government according to market ideologies and structures has been happening for some time. Government budgets are being reduced, and certain public goods are sacrificed so that resources can be used in the more 'efficient' and 'productive' market sector. Yet economic rationalism—also referred to as economic fundamentalism or market dogma—has been soundly criticised on these grounds:

- It relies on narrow mathematical models that oversimplify and do not account for the social and institutional factors that produce variability and unpredictability in human situations.
- It assumes that people are always 'rational', making choices free from social or other factors that might influence their preferences.
- It asserts that markets are always in equilibrium, and it does not account for real and substantial fluctuations.
- It designates the pursuit of *individual* material interests as the optimal way to achieve *societal* well-being.
- It assumes the existence of perfect competition—that no persons or firms can dominate particular markets.
- It makes unsubstantiated claims to be effective in reducing foreign debts.
- It is aggressively promoted and staunchly defended by powerful élites with narrow life and work experiences who are less interested in recognising its limits or hearing alternative perspectives.
- It fails to account for equitable (as opposed to 'efficient') distribution of resources, that is, the gap between rich and poor continues to widen across the world.[3]

Essentially, many are deeply troubled by the global societal and environmental disruption caused by such widespread and rigid adherence to this 'market mentality'.

When there was a downturn in the Australian economy (and in many other Western nations in the 1970s), economic rationalism triggered a series of changes in the dominant ideology of how the public sector ought to be managed. Key élites in Federal and State governments who were trained in neo-classical economics moved to reduce both the size of government and its expenditures and to maximise public sector productivity.[4] These ideals, well engrained in the state bureaucracy by the mid-1980s, were manifest in sweeping public sector reforms, which fundamentally redefined and rearranged how government would operate. These changes were implemented in a quest for greater efficiency and continue to assert the lack of merit in government-owned enterprises, and argue that there should be little difference between how the public and private sectors are managed.

These practices have been referred to as 'managerialism' because managers in different contexts have been the primary means of carrying them out.[5] Like the principles underlying bureaucratic forms of organisation, managerialism reveres 'rationality' and is unswervingly optimistic about how much control can be gained by increasing amounts of organising. Managerialism also asserts that *better* (more efficient) management can be successfully applied to a range of social and economic problems, and that those techniques can be taken from the private sector and applied to the public sector.

Managerialist models of public administration rely heavily on corporate management frameworks. These major tools of this approach include: having clearly defined objectives which management uses to achieve performance levels comparable to the private sector; implementing peformance evaluation to improve accountability; applying market-based systems of rewards and sanctions for directors and management that are closely linked to individual and organisational performance; obtaining managerial autonomy for boards of directors and management in order to achieve objectives; and creating a 'level playing field' by removing any advantages or disadvantages that apply to public corporations by virtue of government ownership.[6]

Some suggest that traditional public management models have outlived their usefulness and corporate reforms are long overdue. They welcome the use of corporate tools in public sector settings such as school administration, because they believe these tools help to provide more flexible organisational structures that operate more efficiently and give the organisation more control in competitive *market* settings. Others are less convinced of the benefits of managerialist reforms for public organisations. It has been asserted that managerialism naively assumes a world-wide solution to common problems, propagates a culture of fear whereby speaking out against management is discouraged, and places too much faith in markets. There are similar concerns about an executive imperialism fostered by managerialist practices and whether these methods really do help organisations to achieve their goals.[7] Indeed, such corporate frameworks may even condemn organisations to poor performance. Some research has shown that efficiency problems often result *from* overly centralised structures, inadequate staff and client involvement in decision-making and from failure to develop open, more problem-oriented policy-making processes—all typical traits of corporatised organisations (see Chapter 5).

Managerialism has particular ramifications for cultural, health, educational and environmental institutions. In effect, the Commonwealth and State governments have retreated from their responsibility for fully supporting these institutions while simultaneously imposing greater use of business principles in the management of them. Des Griffen, former Director of the Australian Museum in Sydney, lamented the Australian governments' administrative policies of 'rational' economics:

> What government has also tended to do is to impose further regulation on government enterprises, but lessen it in respect of commercial enterprises. Governments have tightened regulations about disclosure in annual reporting, introduced accrual accounting, sought to have [museum] collections valued as assets (which misses the point entirely), introduced job evaluation, award restructuring and enterprise bargaining and, in the UK and in NSW, demanded the formulation of performance guarantee statements . . . What government has not done is promote leadership, work out mechanisms to match performance with resource allocation or develop review mechanisms which actually identify what is wrong with organisations that do not seem to be achieving their agreed goals.[8]

Why are these practices so pervasive and enduring when there is substantial theoretical, professional and personal opposition to them? Some suggest that managerialism is so popular because of its strident promises to deliver clear and certain answers. As political and economic environments change and problems become more complex, there seems to be a corresponding rise in people's desires to find clear and simple solutions to these dilemmas. Irrespective of how valuable corporatised management models might be for public bodies delivering identifiable services to individuals, there are certainly doubts about how useful they are in the context of other institutions that have much broader public-good goals.

Managerialism also derives its power from a strong base of institutional support. For example, while Jeff Kennett was the Premier of Victoria virtually no aspect of public policy was left untouched by managerialist reforms.[9] With some of the groundwork for such changes already laid down by the previous State Labour government, the Liberal Party was able to capitalise on economic reform being a top priority on international and national policy agendas. The Kennett government became well known for its rapid and authoritative style of decision-making. It succeeded in substantially reducing the size of the public sector by privatising and contracting out certain of its functions, separating government policy-making from service provision, and introducing contractual relationships between core government departments and service deliverers and—within the public service—between superiors and subordinates. The effect of these changes was to downplay the role of public participation in achieving social goods, to exclude certain targetted interest groups and policy communities from government decision-making, to make government *less* accountable to 'consumers' and 'users' of its services and to favour particular social and economic values.[10]

Within individual organisations, managerialism acts as a 'discourse of the powerful' where financial controls and performance monitoring further a specific agenda of a political, managerial élite.[11] Performance is assessed according to measurable or quantifiable goals or outputs which are defined primarily in terms of economic efficiency. These appraisals are fundamentally the domain of central and higher levels of the organisational hierarchy where the processes of establishing corporate mission statements, goals and directions are controlled. Broad aims are then refined down to smaller action plans, and managers are responsible for ensuring these lower level activities are carried out successfully and advance broader objectives of the whole organisation. These greater demands for reporting and planning, narrow definitions of performance and greater pressures to standardise outputs—trademarks of corporate frameworks—restrict policy. Since activities are cast primarily in terms of outputs, providing discrete goods and services, the full range of values—many of which can only be expressed as intangible—and their complexity are devalued and suppressed.[12] Instead, the managerial system is predominantly concerned with 'markets' for disposal of the 'product' and for 'procurement' of the resources.

A Managerial Ark

These policy trends have significant ramifications for the publicly-owned and managed zoos in New South Wales, Victoria, Western Australia, the Northern Territory and New Zealand, as well as those 'private' zoos in Queensland and South Australia. A corporate model has influenced all Australian zoos, but especially the Western Australian, New South Wales and Victorian zoos. These statutory bodies had a relatively high degree of autonomy until the mid-1980s, but were corporatised by proponents of the managerialist model to bring them under closer ministerial scrutiny and to increase their 'efficiency'.[13] New Zealand has also undertaken an extensive program of corporatisation, privatisation and commercialisation in its public sector. Like its Australian counterparts, the Auckland Zoo, under the governance of the Auckland City Council, has been strongly influenced by a corporate ethic. All these zoos must answer directly to government mandates and policy prescriptions. Hence, they are particularly vulnerable to the institutionalisation of managerialism in the public sector.

The managerial values of private industry are actualised in zoos through an array of practices and techniques. We saw how the senior management strata in certain parts of zoos have grown in recent years. There is now a tangible preference for those positions and zoo board vacancies to be filled by professional bureaucrats or people with business expertise, rather than people with qualifications in zoology, veterinary science, conservation or the social sciences. These trends signal a shift in values, not simply occupational changes. In addition to ministers and senior government officials, zoo board members and CEOs or zoo directors associated with the larger, more corporatised zoos often see their organisation first as a business that must operate 'efficiently' in order to excel in a competitive market. Consequently, private sector 'tools of the trade' are applied to narrowly measuring a zoo's performance, such as budgeting measures that shift notions of accountability into cost-answerability and economic efficiency.

The desire to increase a zoo's efficiency and effectiveness are admirable and not in dispute here. However, there do seem to be questions:

- Are these corporate tools appropriate ways to measure and improve zoo (conservation) performance?
- Does the array of corporate management tools do more to legitimise managerialist frameworks than they do to further zoos' political and social responsibility and environmental reform?

The blurring of means and ends has important ramifications for all zoo policy when the definition and fulfilment of conservation objectives are recast strictly in terms of economically-defined imperatives, objectives and parameters. This discourse has typically been sanctioned and controlled by élites: politicians, senior government officials, zoo board members and zoos' senior managers. It is exemplified by the following passage from a paper given by Chris Larcombe, former Chief Executive Officer of the ZBV and Director of Melbourne Zoo, at an Australasian regional zoo conference in 1994:

> [The zoo community must] come to terms with the increasing financial pressures on the operations of our properties. What we are talking about here is a sustainable base of economic support and leveraging of resources in order to continue to develop our properties in an increasingly complex environment . . . But we will only be able to continue to deliver this [conservation] potential if we have organisations of sustainable financial viability.[14]

This steadfast concern with economic efficiencies is similar to the reasoning infiltrating most environmental administration: neoclassical economic theory dominates thinking about how the environment is best managed.[15] Conservation is something that needs to be integrated with a pro-business approach in order to ensure economic growth.[16] Hence, the free market is increasingly seen as the most efficient means for environmental management (for example, deregulation, privatisation), while (less potent) government programs must increasingly be justified according to economic efficiency criteria. Like other agencies facing shrinking resource pools, zoos have been persuaded to rationalise their conservation activities according to the cost-benefit analysis that is inherent in corporatised management frameworks. Cost-benefit analysis is a tool which decision-makers use to choose between alternative courses of action in deciding whether a project should proceed or not. More specifically, the costs of proceeding are weighed against the benefits that would arise from doing so. For the sake of uniformity, these costs are expressed exclusively in monetary terms. The decision-making tools and value perspectives of these approaches are confining, and can have serious consequences for conservation, as Neil Evernden warns:

> The kind of evaluation permitted by our societal institutions is simply too narrow . . . applying monetary value to nature is dangerous . . . monetary evaluation distracts us from the fact that the values at issue are not economic in the first place . . . and asserts the assumption that human beings are the sole bearers and dispensers of value . . . that one being's existence can be justified only by its utility to another.[17]

The anthropocentric processes that confer economic and commercial value onto non-human nature or efforts to conserve it give zoo conservation policy a particular character. That is, official pressures and budget constraints balance conservation against other demands. Conservation values translate into activities—not something that the zoo (or any other institution) *must* do—but something that the zoo (or any other institution) must be able to *afford* to do. Using cost-benefit analysis as the sole or main decision-making tool 'resembles the proverbial carpenter whose only tool is a hammer, and who sees every problem as a nail'.[18] Yet environmental problems are different—conservation has worth that far outweighs monetary value. If one wanted to calculate the full price of conservation, *all* the *non-economic* benefits associated with it would need to be included. Barring the fact that we have not found sufficient methods for placing monetary value on non-human nature, any price that is calculated will be far more than what we can pay. Institutions, organisations, and individuals can then always claim to be doing only what they can afford to, and no more.

No doubt, modern zoos face formidable challenges in maintaining some financial viability in a political and economic climate that threatens all cultural, educational and recreational institutions. The constant struggle to resource public goods, such as conservation and education (which have worth that extends well beyond monetary value) is something that troubles many members of the zoo community. However, zoo professionals understand the underlying context of their institution's situation in a variety of ways. Zoo staff's frustrations were examined earlier in this book. Most zoo staff appreciate that some economic imperatives must be addressed, but some are also deeply troubled by what they see as overly bureaucratised and commercialised policies that compromise zoos' public good responsibilities. The presence of these sentiments alone does not necessarily prove that conservation and education values are being compromised, but it does intimate that economic management perspectives and imperatives do not receive widespread and unconditional support throughout the zoos.

It is important to appreciate that senior managers have the unenviable responsibility of responding to certain political and economic pressures that staff working in the lower levels of an organisation do not have to deal with directly. Technical staff (for example, animal keepers, horticultural staff, and zoo educators) are largely limited to working within their immediate, internal organisational environment. Senior managers in zoos must often procure needed resources, mediate between the organisation and its customers and act in a leadership capacity. Moreover, with the wave of managerialist reforms, many senior executives are now under *contractual* obligations to show what outputs they are achieving. Caught between the mandates of their controlling boards of directors or government officials and their zoo staff, senior managers are often referred to, or refer to themselves, as 'the meat in the sandwich'. They must reconcile the interests and demands of those above and below them and are ultimately answerable for their organisation's performance.

The official pressure on public zoos, in particular, to justify their existence and the scope of their operations according to private sector criteria is real and has been increasing. Under the guise of accountability, publicly-funded zoos in Australasia must regularly report back to their respective State governments about whether organisational goals have been reached. Zoo professionals use plans and performance measures to illustrate to themselves and their governmental masters how efficient and effective their organisations are. Zoo professionals must also account for the 'true' costs of their organisations by charging for services of various departments, market rents and lost opportunity costs to its user departments. Is there not some danger that conservation imperatives—which have no 'economic' return *per se* and are difficult to quantify—will be lost under a mandate to 'manage' zoos in such a way that costs are minimised and revenues are maximised?

Defining the Ark's Business

Zoo professionals use an array of corporate strategies and planning documents to provide comprehensive and proficient assessment of a zoo's operating environment, and to identify opportunities, threats, market niches and appropriate revenue mixes. Collectively, these corporate tools also provide formidable support for managerial values, but their evaluations are only partial given their limited focus on the *commercial* aspects of a zoo's operations.

Under the instructions of the Auckland City Council, the Auckland Zoo undertook an extensive strategic planning exercise in 1992 to consider its performance and future opportunities and challenges. This work included a comprehensive Strategic Plan for 1995/96. The Plan listed several factors contributing the Zoo's changing (and challenging) operational environment including, 'greatly increased competition in the entertainment and recreation markets, increased expectations of service and value for money, increased questioning of the relevance of zoos, decreasing government spending and increased costs'.[19] This policy document was slightly different from others like it, because it distinguished nature conservation, not as a marketable 'product', but as an 'outcome' of the range of different Zoo programs. The 'marketable' product was identified as 'an enjoyable, memorable, visitor experience'. Nonetheless, the strategy relied on developing an adaptable, service-oriented organisational culture that would enable the Zoo to survive in tough economic times:

> [The Zoo *must*] develop an infrastructure and a mentality that is customer focused and adapts quickly to change, including the ability to respond quickly to revenue-generating opportunities . . . it must provide experiences every year which are fresh, exciting, and attractive in the entertainment and recreation marketplace at a competitive price.[20]

This perspective is not unique to Auckland Zoo administrators and managers. Some members of the Australasian zoo community feel a zoo must be able to compete on the same terms as the rest of the 'recreation business'(see Chapter 4). This line of thinking means that zoos have to be able to offer escalating levels of stimulation in order to keep first-time or repeat visitors from casting zoo visits aside in favour of other, more extraordinary options. There are significant policy and fiscal ramifications that flow from this standpoint.

First, this perspective devalues what a zoo already has on offer. Watching animals is assumed to be less appealing than what is offered by other recreational venues. Second, the idea that zoos must always offer new (and exciting) exhibits and attractions locks an organisation into a particular spending cycle, where substantial funds are required to pay for the continual development of multi-million dollar exhibits or to maintain marketing and implementation infrastructure for special events. Moreover, while new exhibits and events help raise admission figures, these increases are typically temporary and admissions drop off after the newness of an exhibit wears away. The sustainability of these practices seems tenuous. Moreover, they could merely be reinforcing—instead of changing—

public expectations that a zoo should be like other recreational and entertainment venues.

This kind of logic also appeared in the draft version of Perth Zoo's five-year business plan for 1993–1998. The Master Plan component of the strategy assumed that the principal objective for further physical developments of the Zoo should be to maximise revenues. Again, 'new' is assumed to be 'better.' A business strategy was articulated, and critical success factors and key objectives were identified which the management team would be responsible for achieving in their respective divisions. The Plan also contained a ten-year milestone wish list, a stakeholder analysis which lists the degree of power and influence of those groups and what they want. Incorporated into the Plan were an environmental scan of the latest social, legal and political trends; a list of opportunities for and threats to the Zoo; a customer and competitor analysis; and the Zoo's competitive, financial, technical and human resource strengths and weaknesses.

The document's corporate and commercial imperatives were readily apparent. It was also made clear how problematic this kind of planning can be when means overshadow ends. The Plan identified 'too much preoccupation with planning and restructuring and too little action towards achieving the Zoo's objectives', 'too much paperwork arising from accountability requirements', and 'a greater focus on economic issues and less on conservation' as *significant* threats.[21] Despite these insights, this and subsequent plans have been dominated by corporate and commercial imperatives. For example, these are the draft Plan's 'Highest Priorities':

- to remain high in public profile and maintain a favourable public image and become commercially oriented in its Marketing and Public Relations;
- to become and remain customer focused and assure customer satisfaction;
- to continue to acquire sufficient resources (particularly financial resources) to achieve the above and fulfil its mission;
- to remain easily accessible to its customers in terms of location, entry points and hours of opening.

The need for these actions was certainly portrayed as indisputable. The Plan identified a significant lack of marketing resources and culture as needing attention, consistent with the view 'that the Zoo, although it is a conservation organisation, is essentially in the Recreation business', and that by 'increasing commercialisation and privatisation' the Zoo would be able to gain 'access to both Government and private sector funding'.[22] A stronger commercial orientation was clearly positioned in the Plan as the remedy for the Zoo's ills and was certainly consistent with the outcomes of the Zoo's subsequent organisational restructure (see Chapter 5).

By 1995 another Business Plan for a five-year period was in place, endorsed by the Zoo Board and implemented by the new management regime at Perth Zoo. The new plan made no mention of the kind of concerns registered by zoo staff in the earlier draft business plan for the period 1993–1998. It reiterated the fundamental concerns of the previous plan and stated again that 'whilst the zoo is a conservation institution, its main operations are essentially in the recreation business'.[23] Implementing the new Plan would supposedly achieve these goals:

> ensure the Zoo's long-term viability by becoming increasingly self-sufficient and independent of government funding, saving taxpayer's money, increasing the range and quality of services delivered to the community, whilst realising our conservation mission, fulfilling our community service obligations, and achieving our vision by implementing our Master Plan cost effectively.[24]

The viability and cost-effectiveness of this highly ambitious Plan could be debated. In effect, it actually locked the Zoo into a dubious spending cycle predicated on the notion that continual development of new exhibits can achieve sustainable increases in visitor admissions, and that decreased State government spending on cultural institutions is appropriate. The Zoo and its Board were engaged in an 'essential' capital works improvement program costing approximately $24 million dollars. Half of that amount was to be borrowed from the State government, in turn the Zoo promising to pay for its *own* capital development after 1999/2000. Private sector sponsorship would pay both the debt incurred and the second half of the $24 million. In addition, the Zoo promised its government that the current ratio of government support—65 per cent—would be improved to only 35 per cent by 1999/2000. Furthermore, the recurring government grant would be reduced from 66 per cent of revenue to 25 per cent by 2004/05.

These measures (and others like them) point to some zoo administrators' industrious attitudes, entrepreneurial creativity and willingness to co-operate with government policies. Their degree of optimism may not be justified. There are consequences that flow from using corporate tools as the first point of departure in defining the 'business' of zoos. Departments with budgeting powers and revenue-raising capacities must seek out revenues and ensure costs are controlled. They are, in effect, placed on the offensive while other program areas operate more defensively, having increasing difficulties justifying their expenditures or developing some ability to raise money. A culture of cutting costs is further entrenched when continual strains are placed on current revenue streams from plans for frequent, large capital works improvement plans. This is not meant to diminish the importance of improving the living conditions of animals when existing exhibits are substantially upgraded. Indeed, this ought to be a high priority for zoo developments. What seem less practical are projects which feature new, sensational, high-profile species in the hopes of attracting ever more visitors. A case in point might be the McDonald's Gorilla Forest at Taronga Zoo, which cost the Zoo nearly $4 million to build.[25] A sizeable proportion of this money was supplied by the corporate sponsor, which is listed by the Zoo as a 'Crown Sponsor', donating over $1 million in cash and/or in-kind support.[26] Yet the idea that zoos will be able to collect an ever-increasing proportion of their income from the private sector for projects like this may not be realistic. There is already intensive competition for the corporate and philanthropic dollar. Some cities, such as Adelaide and Perth, are further disadvantaged because there are not as many corporations' headquarters located there, as opposed to metropolitan centres like Sydney and Melbourne.

Efficient Management of Ark Passengers and Manifest

We saw in Chapter 5, zoos are positioned as part of the service industry. A product form of organisation is used to arrange zoo activities into different areas. These programs are then rationalised according to their operational costs and performance, and their achievements are expressed as product-like entities. It is valuable and necessary to delineate work activities and evaluate their effectiveness. Certainly zoo administrators can demonstrate some level of accountability to their government masters by using these tools. Yet, in a sense, the measures managerialism endorses are more proficient at defending the corporate infrastructure that implements them, than they are at illustrating the tangible *quality* of the *whole range* of zoo activities.

An overemphasis on outputs disregards the fact that some qualitative, organisational values are less tangible and therefore defy 'product structuring'. Quantitative economic measures imposed on zoos put animal and conservation-related objectives in competition with recreational and commercial objectives: the ones that bring in the most money gain the most support. Similar concerns have been raised about the use of corporate tools to manage universities. An emphasis on matching departmental or individual outputs to economic prescriptions excludes cultural and academic values. Moreover, power is conferred exclusively on senior managers who must identify which divisions and/or people are responsible for incurring profits and losses.

The notion that regular, quantitative measures of an organisation's performance can help it achieve its desired outcomes is manifest more strongly or openly in some zoos than in others. The Victorian Zoo Board has traditionally not reported publicly on its performance measures. The Adelaide Zoo has managed to avoid using extensive performance indicator systems. It provides general reports about issues such as visitor attendance trends, what conservation programs are operating, corporate sponsors and standard financial statements. The Western Australian Zoological Board and the ZPB of NSW have made the most obvious and public use of formal performance measures in the last decade (see Tables 6 and 7).

Table 6 Quantitative indicators used by Perth Zoo

Effectiveness measures	**Performance indicators**
Wildlife conservation	• number of species kept at Zoo involved in conservation breeding • number of viable offspring produced as result of breeding program for reintroduction • number of endangered animals provided for release into wild as part of collaborative conservation programs • number of endangered species breeding management plans completed
Customer awareness	
Conservation, education of visitors and attitude changes	• percentage of visitors agreeing with 5 conservation education questions about the Zoo
Communication of conservation message	• percentage of visitors agreeing that the Zoo's conservation message was communicated to them
Zoo customers	• number of total admissions to the Zoo • frequency of visitation
Customer satisfaction	• percentage of visitors supportive of quality of animal presentation, courtesy of zoo staff and quality of zoo services
Efficiency measures	
	• proportion of Zoo income to government appropriation • expenditure per visitor admission • income per visitor admission • admissions per full-time staff

SOURCE: Compiled from Perth Zoo, *Annual Report,* 1993/1994, 1997/1998.

By the early part of the 1990s, Perth Zoo was reporting on its quantitative and monetary indicators to measure its success. For the 1993/94 *Annual Report*, the Zoo was using quantitative indicators to measure its effectiveness and efficiency in achieving a number of objectives. By 1998, the indicators were largely the same, but some significant shifts in emphasis could be detected. Reporting on efficiency indicators was more extensive and now included the *costs* of operating the *education and wildlife conservation* services.

Table 7 Indicators and outcomes for Taronga Zoo and Western Plains Zoo

Performance indicators	Taronga Zoo	Western Plains Zoo
Increase in endangered species represented in collection	0.9%	1.7%
Increase in Australasian species represented in collection	3.0%	n/a
Increase in number of students visiting the zoos	1.9%	17.5%
Increase in retail sales	7.5%	1.1%
Estimated value of free media coverage	$3 810 568	$673 860
Number of days of fieldwork completed on Platypus research project	420	n/a
Number of scientific studies undertaken	16	n/a

SOURCE: Compiled from Zoological Parks Board of NSW, *Annual Report*, 1996/1997.

There are several questions worth asking here. First, what are we saying about the value of animals when we assign dollar costs to maintaining each species or specimen and quantitatively measure their contributions to breeding programs? It is as if an animal or species' inherent worth is in direct competition with its utility value: an animal is valuable insofar as it is deemed 'useful' (for example, known to attract more visitors). While it may be helpful to know the *costs* of maintaining animals, this is not the same as knowing the full *worth* of animals. When one considers this kind of logic, it is not entirely surprising that cynics have portrayed an economist as someone who knows the cost of everything but the value of nothing. Second, not all program areas at Perth Zoo appear to be subject to the same kind of corporate and public scrutiny regarding their 'effectiveness' and 'efficiency'. The fact that Education services were being assessed in the annual report in terms of their *costs to the Zoo* provides further evidence in support of suggestions made in earlier chapters that the power bases of zoos' education programs are being supplanted by a marketing, customer-service orientation. These instrumental and commercial means of assessing the worth of animal management and education programs increase government officials', boards of directors' and select senior managers' ability to control a zoo. It is not entirely obvious how they increase a zoo's capacity to deliver quality education or conservation programs.

Quantitative performance measurement also occurs in NSW, although these tools vary in how they are listed in the Zoological Parks Boards' annual reports. In the financial year 1992/93 the Zoo Board reported that it had established a comprehensive mechanism for performance evaluation. Unlike its Western Australian counterpart, the Zoo Board's annual reports in subsequent years include performance indicators for *all* its program areas. Nonetheless, these indicators portray Taronga and Western Plains Zoos largely in a favourable light and are primarily quantitative. These numerical and monetary representations of the Zoos' program performance are impressive and demonstrate a particular kind of success. They do not speak to other forms of knowledge and insight. The example of how zoos can value (favourable) media coverage is a case in point. Some zoos, like the statutory authorities in NSW, Victoria and WA, assess the worth of media exposure by calculating what it would have cost them in paid advertising time. These figures are then used as performance indicators to demonstrate the merit of their public relations or marketing efforts. Here 'more' of something is equal to 'better', especially when it is free. These monetary indicators of success do not speak to the *quality* of that media coverage. Promotional exercises have the potential to mislead the public and compromise the educational integrity of the zoo by overstating the zoo's role in conservation or by discouraging controversy and public debate (see Chapters 4 and 5).

The ZPB of NSW's benchmarking initiative is another example of quantitative corporate tools for measuring and improving performance. The benchmarking process requires two or more institutions to share (quantitative) corporate information in order to compare their performance in particular areas with other organisations, in order to determine which practices are 'best' and worthy of emulation. Modeled after Best Practice Programs in the manufacturing and agriculture industries, benchmarking systems are endorsed by industry leaders who perceive a need for widespread change in organisational cultures in order for Australasians to become more competitive in the global marketplace. Glenn Smith, Deputy Director of the Zoological Parks Board of NSW, stated:

> Organisations operating in competitive industries such as mining, retail and manufacturing have long been searching for methods to improve performance and productivity. Zoos have also been striving to improve performance in a wide range of areas from species conservation to customer service and commercial viability. An important management tool used by competitive organisations is the measurement of performance aimed at achieving industry best practice on both a national and international level . . . the process of benchmarking will help to crystallise standards of perfect performance (not merely current best practice, but moving beyond to the best possible level of practice) towards which an organisation can aim.[27]

Zoo professionals wishing to raise their performance standards and problem-solving standards should be commended. What remains debatable, however, is the perspective that zoos are an 'industry' whose community should behave accordingly. For example, while sharing information is stressed by benchmarking, this system might foster more competitiveness—not co-operation—between institutions. Benchmarking assumes that 'where an activity is carried out by two or more groups of people, it is normal that one group will achieve the task in a better fashion than others'.[28] The chase to be the best 'becomes the benchmark, but will be overtaken as others improve their operations'.[29] Competitiveness seems to be openly and uncritically embraced. Social commentators have warned that overall, competition as an operating societal (or business) principle is harmful and suggest that 'civic virtues come from building on what we have in common rather than using our differences to create in-groups, outgroups and fear-driven competition'.[30] Furthermore, what has not been openly clarified by zoo or other benchmarking systems is *who* decides what constitutes 'better', who it is supposedly 'better' *for* and whether it is appropriate—or even possible—to measure all aspects of an organisation's performance on the basis of quantifiable indicators.

These quantitative procedures are informed by the same kind of positivist frameworks that have weakened zoo research. Benchmarking measures such as the number of visitors per employee per annum, the total annual operating cost of zoos divided by total specimens held or the number of specimens per zoo employee, reduce (the worth of) people and animals to numbers. Further, it is considered valid to break down complex phenomena into smaller and more manageable pieces and to analyse them separately. In some cases, this can be useful. Certainly, quantitative knowledge lends itself more readily to measurement. But it is not correct to assume that this kind of data is more valid and useful than qualitative information, which cannot be determined in the same way. If quantitative studies of variables are consistently chosen in preference to qualitative studies, a very partial—rather than holistic—picture emerges. Quantitative technical information is not sufficient for understanding and addressing conservation problems (which are embedded in complex social contexts), nor can it fully represent the true worth of conservation.

Where there is a genuine interest in understanding people's (or animals'!) lives, stories and behaviours, qualitative assessments will be more appropriate for resolving certain organisational issues or evaluating the particular activities of varied program areas. The impact of educational programs on visitors' environmental behaviours or a zoo's organisational climate are two examples. Qualitative inquiries can provide novel, fresh perspectives on matters that cannot be captured by statistical procedures.

How Do We Pay for the Ark?

The way zoos are financed is another important influence on their policies. The increasing influence of economic rationalism has reduced government support for museums, universities, libraries, health and environmental organisations and public zoos. These institutions are being told that, apart from support for some community service obligations, they need to become more efficient and economically viable by reducing their dependence on 'government handouts' and more fully developing their commercial potential. It would be very unrealistic to suggest that, in this frugal climate, zoos and other cultural and health institutions should not be concerned with maximising and stabilising their revenues. Most public or private organisations seek out stable growth, decision-making autonomy, and control. The more relevant concerns are about how much responsibility governments *should have* for supporting cultural organisations providing a public good and to what degree certain financial policies and planning procedures have a negative impact on these institutions' ability to provide such public goods.

Table 8 shows that, overall, zoological parks and aquaria have had a greater commercial capacity than other similar cultural institutions, drawing approximately 18 per cent of their income from government, while nearly all of the remaining revenue has been derived from commercial activities (including sales/trading, admissions, private sector income). Table 9 illustrates that in the late 1990s the amount of government support for Australia's four major zoos has actually been higher than the industry average, with zoos in WA and SA relying more heavily on government support, while their counterparts in NSW and Victoria have strong commercial programs. Certainly admissions income remains a critical source of revenue for all zoos. Tables 10, 11, 12 and 13 demonstrate that, during the last ten years, State government support of these major zoos has dropped while commercial income has grown. Not surprisingly, this trend is more notable in the statutory zoos, which are more readily influenced by State government priorities. This is particularly true for Perth Zoo, which has historically drawn the greatest proportion of its income from government. Interestingly, the fluctuations have been less notable for Adelaide Zoo and Monarto Zoological Park in SA. The Royal Zoological Society of SA, which manages the two zoos, has been able to operate more independently from government, even though it has obtained the highest proportion of its income from government. The smaller size

Table 8 Some cultural institutions and sources of income

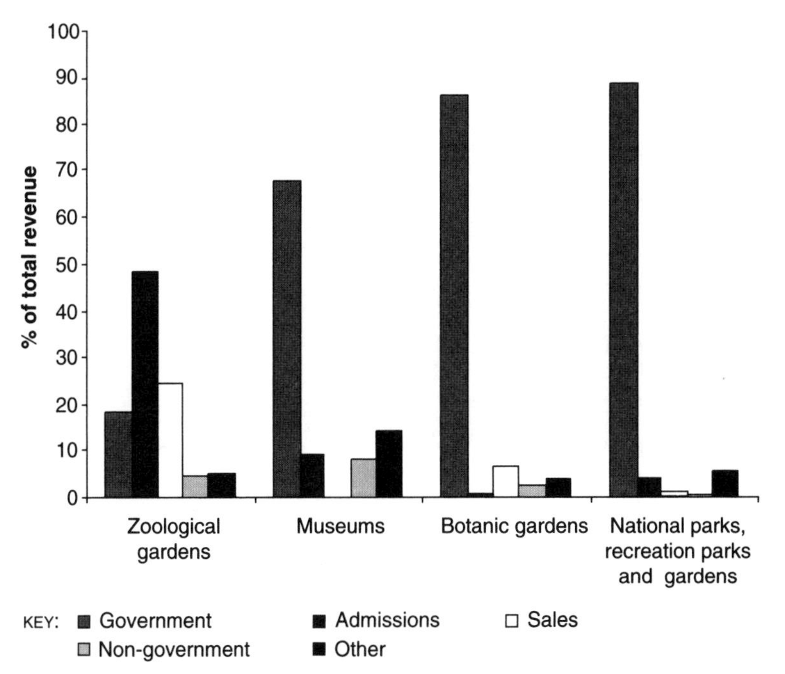

SOURCE: Compiled from Australian Bureau of Statistics, *Zoos, Parks & Gardens Industry 1996–1997*, Catalogue no. 8699.0, pp. 7, 13, 19; *Selected Museums 1997–1998*, Report no. 4145.0, p. 10.

of the Society's two zoos (relative to their interstate counterparts) and the Society's culture, which is steeped in zoology and conservation, may partly explain its ability to be more resistant to economic rationalism.

Despite these variations, most zoo professionals are still attempting to improve their institution's commercial capacities and develop greater infrastructure for securing private sector funding, both to top up shrinking (or stabilised) government assistance and to act as an insurance policy for funding special projects and continual capital works improvements. After-hours or 'value-added' events are one relatively new commercial mechanism used at zoos. These programs, designed to increase admissions incomes, are

Table 9 Current sources of funding* for four major Australian zoos

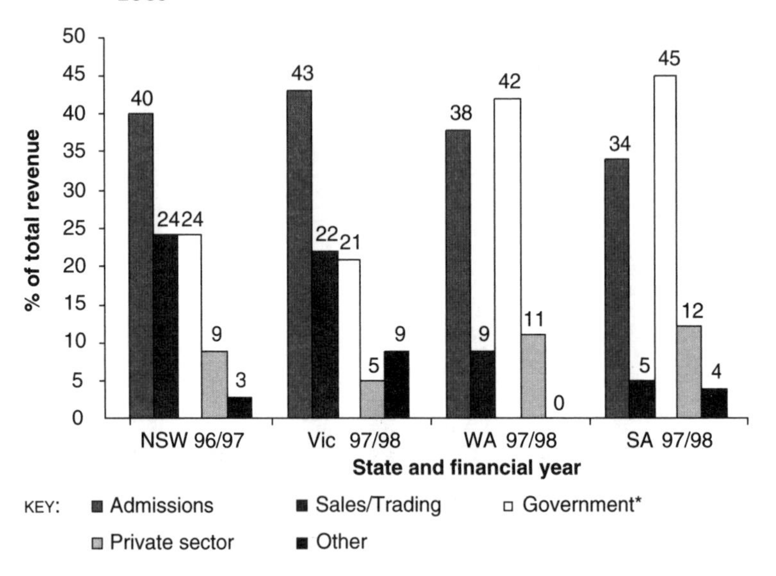

* Includes capital funding

SOURCE: Zoological Parks Board of NSW, *Annual Report*, 1996/1997; Zoological Parks and Gardens Board of Victoria, *Annual Report*, 1997/1998; Zoological Board of Western Australia, *Annual Report*, 1997/1998; Royal Zoological Society of South Australia, *Annual Report*, 1997/1998.

typically conducted during the summer season and include after-hours tours, concerts, and corporate evenings. The Zoological Parks Board of Victoria offers a selection of special programs at its three properties. During its Zoo Twilights series, Melbourne Zoo extends its opening hours to nine o'clock at night and provides jazz concerts, a barbecue, and special meet-the-keeper sessions. At Healesville Sanctuary and Victoria's Open Range Zoo (VORZ), the after- hours programs are designed to be consistent with the theme of each property. Given its emphasis on African grasslands, the VORZ after-hours concerts feature live African music, African food, and African art and craft activities for children, in addition to the Zoo's

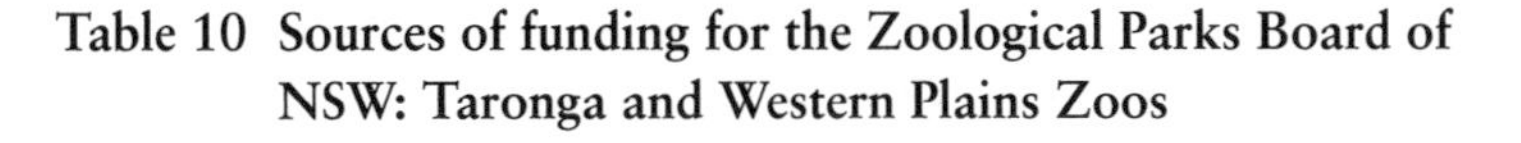

Table 10 Sources of funding for the Zoological Parks Board of NSW: Taronga and Western Plains Zoos

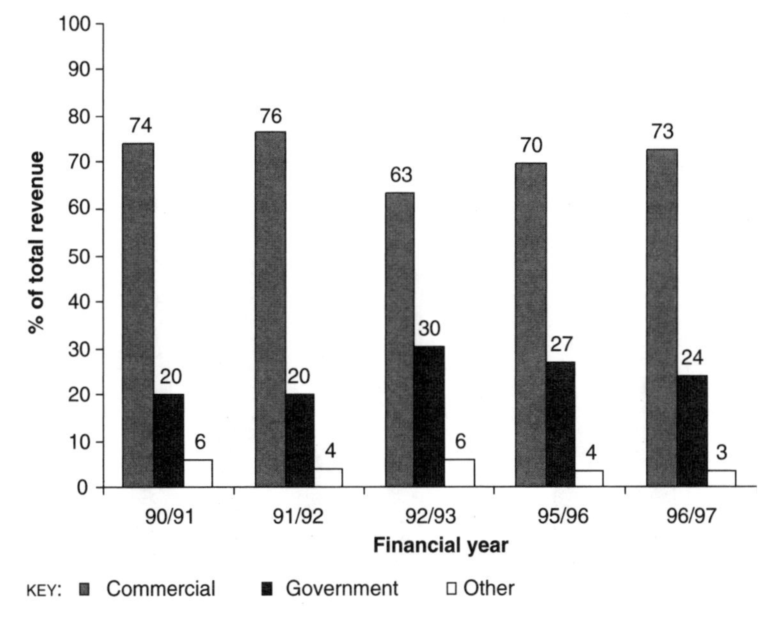

SOURCE: Zoological Parks Board of NSW, *Annual Report*, 1990/1991–1996/1997.

wildlife tours. Healesville Sanctuary's program features rock/acoustic concerts and is called Unplugged—species survival series.

These events provide a convivial and enjoyable evening for the general public. They are also popular. Admissions figures for after-hours events and corporate after-hours functions totaled nearly 146 000, an estimated 11 per cent of the Zoo Board's total admissions for 1997/98. Are the revenues raised by these kinds of programs greater than the costs required to implement them? To what extent do these events further zoos' conservation and education missions? By becoming evermore adept at implementing these programs, do zoos risk shifting their evolution away from the trajectory towards 'environmental resource centres' back to purely recreational and entertainment venues?

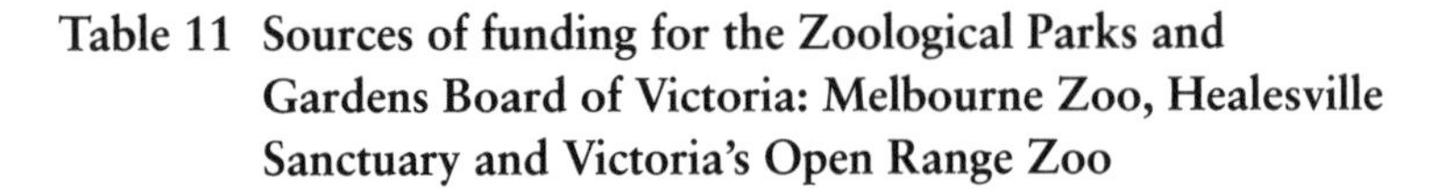

Table 11 Sources of funding for the Zoological Parks and Gardens Board of Victoria: Melbourne Zoo, Healesville Sanctuary and Victoria's Open Range Zoo

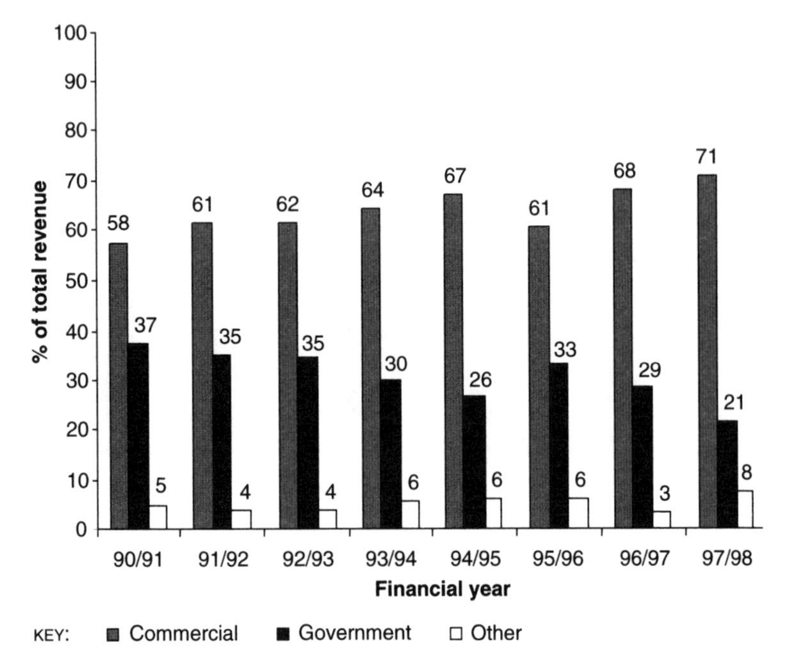

SOURCE: Zoological Parks and Gardens Board of Victoria, *Annual Report*, 1990/1991–1997/1998.

It seems likely that events, such as the concerts at Healesville Sanctuary with a specific focus or special theme, may be more capable of delivering conservation or education messages. Advertisements and publicity for the series 'Unplugged' state that proceeds go to the Sanctuary's endangered species breeding programs. The zoo staff member introducing each concert's performer reiterates this and speaks briefly about the Sanctuary's conservation programs. It is unclear whether this information will actually inspire people to visit the Sanctuary more often and/or to engage in environmentally responsible behaviours.

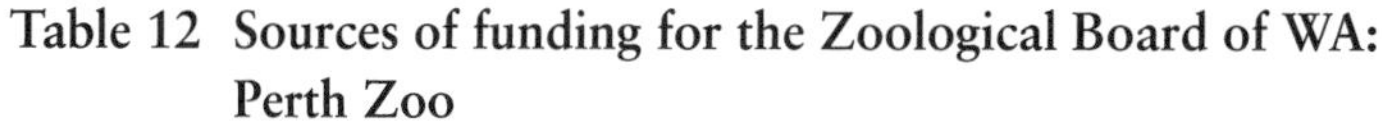

Table 12 Sources of funding for the Zoological Board of WA: Perth Zoo

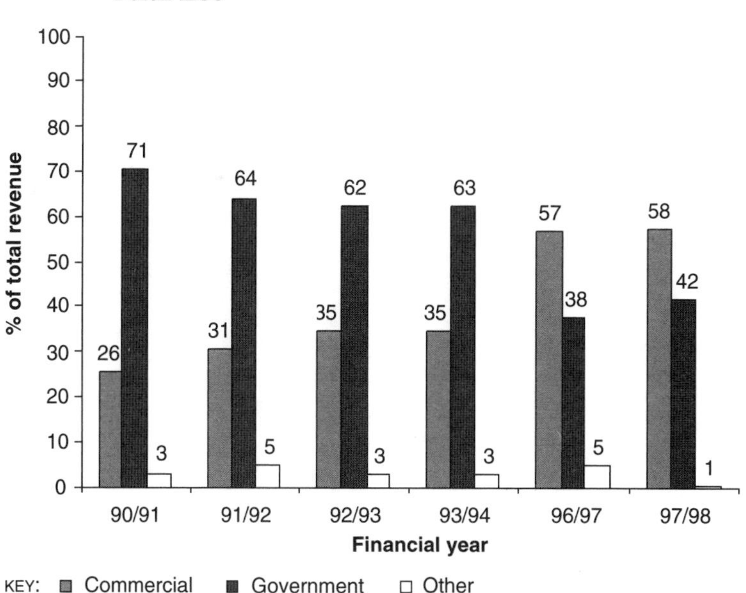

SOURCE: Zoological Board of Western Australia, *Annual Report*, 1990/1991–1993/1994, 1996/1997, 1997/1998.

There is also some ambiguity around the full cost of these programs and how the funds they raised are allocated. A certain amount of infrastructure is required to implement a full program of after-hours or corporate events. Zoo marketing staff must find time in their busy schedules to obtain the substantial amount of corporate support necessary for absorbing the cash costs and/or providing in-kind assistance (publicity, equipment, etc.) for the events. Some controversy has erupted over the pressures these special events place on zoo staff. At Perth Zoo and Melbourne Zoo, there has been some resistance from animal-keeping staff to working night hours and seven to ten day shifts. Some of these staff are unhappy about both the stresses that working longer hours places on them and the use of enterprise bargaining by zoo administrators to ensure that organisational reforms like regular scheduling of after-hours events are realised.

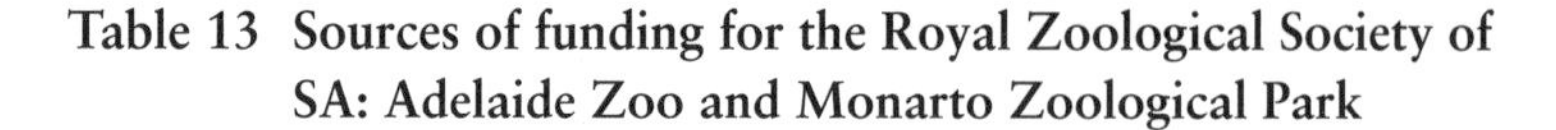

Table 13 Sources of funding for the Royal Zoological Society of SA: Adelaide Zoo and Monarto Zoological Park

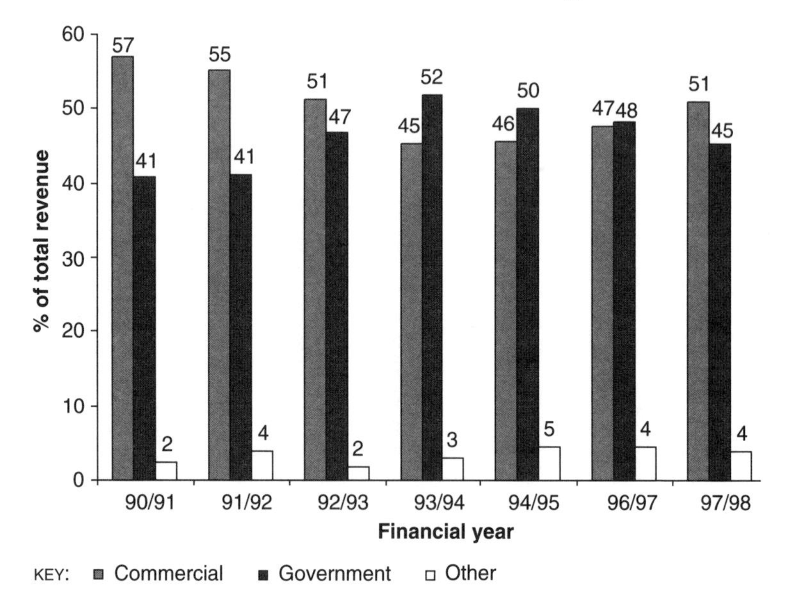

SOURCE: Royal Zoological Society of SA, *Annual Report*, 1990/1991–1994/1995, 1996/1997, 1997/1998.

It is difficult to measure the exact contribution after-hours events make to a zoo's overall income stream, because the revenue from these events—relative to the expense of running them—has not typically been listed in zoos' annual reports. Zoos' (public) financial statements tend to report general financial results rather than costs and revenue from specific programs. Isolating what percentage of a zoo's funds is spent on conservation for that matter would be just as challenging as trying to determine the value of commercial events like after-hours programs. Conservation is defined in varied ways by different organisations and professionals, and may occur in a variety of activities across any given zoo. Even if one were able to sequester the necessary information, it is not clear that such quantitative information would be an appropriate way to value conservation programs which have a greater good than any economic indicator can demonstrate.

Selling a Piece of the (Public?) Ark

Questions about how zoos secure and then allocate revenues and what impact that has on their missions can also be asked of corporate sponsorship. Zoos use corporate sponsorship to fund various projects. The actual donations come to zoos in the form of goods and services and cash, and are typically used to support capital works improvements. Companies assist with building new exhibits, upgrading other zoo facilities, promotions and special events, transport for animals and staff, accounting and auditing services, special research projects and veterinary supplies. Overall, the different forms of corporate support enable zoos to carry out projects they might not otherwise be able to undertake if they had to rely solely on income from their admissions and government grants and subsidies.

Corporate sponsorship provides tangible support for zoos and it also has certain consequences for zoo policy. Despite corporate sponsorship's philanthropic potential, it is still predicated on formal and informal exchanges of benefits. Sponsorship is primarily a market-oriented approach used to enhance the image of a corporation and/or its products. Sponsorship can be used to gain the attention and interest of the public who might have limited interest in a corporation or in that corporation's products. Sponsorship objectives include obtaining some type of favourable media coverage, conveying particular messages and/or increasing sales.

Corporate support for zoos can be linked to what motivates businesses to sponsor environmental projects. Business has responded in particular ways to being identified as a significant cause of environmental degradation. Several categories of business' attitudes towards environmental concern have been identified.[31] *Rejectionists* completely dismiss the need for environmental reform. They typically seek to protect business against what is seen to be unjustifiable attacks and argue that whatever damage is imposed is justified by the economic development and/or growth that results. The mainstream position is taken by *accommodationists* who are skeptical of environmentalism, but seek to maintain as much of the rejectionists' ground while partially accommodating key environmental concerns. For instance, accommodationists would endorse minor, incremental policy shifts in order to conform to sustainable or ecologically sustainable development. Finally, the less common position is taken by *environmental businesses*. Here the need for environmental reform is fully embraced and organisations and product designs are extensively changed to minimise environmental damage.

When corporations sponsor environmental or conservation projects they are able to publicly display their putative interest in environmental issues. Any of the three standpoints discussed above could be motivating business in these instances. Some corporations have a genuine interest in contributing to environmental problem-solving and lend financial or in-kind support to worthy projects. Others—*rejectionists* and *accommodationists*—are wary of further government regulation and/or consumer rejection and look for ways to fend off interference and criticism and to sustain their market presence. In addition to their specific attitudes towards environmental issues, corporate sponsors are faced with increasing levels of 'clutter' in the sponsorship market and are seeking 'new areas of social intrusion with which to associate themselves'.[32] In the United States corporations have spent as much as $254 million on cause-related marketing, $50 million of which has been spent on environmental causes.[33] Cause-related marketing that ties into wildlife and nature is a significant growth area targeted by corporations and has been dubbed 'eco-public relations' given a lack of clarity about the sincerity of businesses using it.[34]

The corporate community does not limit itself to supporting environmental projects when attempting to construct a socially conscious image for itself. Many businesses have traditionally favoured arts sponsorship for achieving objectives relating to community relations and for reaching opinion leaders. Corporate sponsorship of the arts has generated concerns within the arts community. As early as 1986, the Australia Council for the Arts was researching the potential threats imposed on arts programs by corporate sponsorship. The Council found survey respondents were concerned about the following possible consequences: a loss of artistic control and freedom; ethical conflicts; conflicts of values; detrimental effects on the acceptance of responsibility by government and the size of their subsidy; negative impacts on long-term security of funds, especially if private support became a major proportion of income; pressure for popular or commercial success; and strengthening the view of art as a 'commodity'.

It is acknowledged in the museum community that a considerable effort is required to obtain sponsorship from the corporate community.[35] How might the process of selling a 'product' to the corporate community facilitate subtle shifts in *zoo* culture and policy? Under the pressure of rising costs and government policies dictating reduced funding, zoos are seeking to increase their private sector income in general, and corporate sponsorship in particular. We might look at the amount of time, resources and personnel utilised by an organisation to obtain corporate sponsors relative to what is devoted to its other operations. Again, an ambiguous style of financial reporting by public zoos and their relatively protective stance against outside interests makes it difficult to analyse specifically how many resources zoos devote to securing corporate sponsorship. However, it seems reasonable to suggest that if zoos are going to step up their solicitation of private sector funds, more time and resources will be directed to developing further their already sophisticated sponsorship divisions. Chapter 5 provided evidence that some zoos' organisational arrangements have been altered to accommodate this kind of fundraising. The cost effectiveness of such measures is uncertain. An appeal to raise money for Taronga Zoo's rainforest exhibit in the early 1990s secured $119 000 to offset construction costs. Developing the fundraising scheme is reported to have cost the Zoo $47 726 in consultant's fees.[36] This amount does not include time taken by Zoo staff in planning such an exercise. Given the need for funding education and conservation programs, such spending seems to exemplify how commercial means (building new exhibits to attract more visitor revenues) threaten to overtake the ends (achieving species restoration or public education).

The movie *Fierce Creatures* was not critically acclaimed, but the caricatures it used to portray corporate sponsorship run amuck did raise issues that concern real zoo professionals. Parts of the zoo community worry about the practical and ethical implications of relying too heavily on corporate sponsorship to fund exhibits that primarily feature charismatic mega-fauna. Random development and faulty workmanship may be a consequence of planning exhibits around the acquisition of corporate sponsors; inadequate briefing between consulting architects and zoo staff; implementing inappropriate deadlines for completing construction; and depending on media fanfare to announce the opening of new exhibits. One team of zoo architects admitted that most of the major exhibit projects they work on are undertaken because a corporate sponsor is interested in a particular project. The project may or may not be part of the Zoo's Master Plan. Subsequent mismatches between exhibit design, function and animals' needs can result from last-minute planning, tight deadlines for completing projects and a lack of continuity given high turnovers in zoo staff and outside consultants.

In addition to these practical challenges, some zoo staff believe the presence of so many sponsors creates an overly commercial atmosphere and are uncomfortable with animals being 'sold' as products to potential sponsors. Certainly some rather vexing issues flow from seeing animals in terms of what commercial images they can offer for the sponsor. Animals are then objectified and become commodities, rather than existing in their own right, independent of human values. In turn, these messages contradict the ecological ideals that environmental educators hope to convey to zoo visitors and the general community—that non-human nature is worthy of our respect and should be treated with due care.

What other kinds of policy biases might be introduced or reinforced by commercial activities such as corporate sponsorship? In an atmosphere of increasing competition for the private sector dollar, zoos will need to offer something desirable to companies looking to spend their sponsorship dollars. Zoos' choice of conservation programs or exhibit developments might be influenced by this means of support, because corporations (either *rejectionists* or *accommodationists*) will favour projects designed to deflect any negative images the public may have of them. This has certainly been the situation in arts funding. A recent Arts Council survey of businesses contributing to the arts found that 63 per cent of respondents felt that one of several very important reasons for doing so was to improve the image of the company in the eyes of the public. In addition, 74 per cent of respondents agreed that in today's competitive environment, business would only support the arts if direct commercial benefits were received in return.

Given the view of many marketing professionals in zoos that conservation can 'be hard to sell', it seems likely that corporate support for high-profile, charismatic species (or for ecosystems such as rainforests) will be seen as easier to obtain than securing support for obscure, lesser known species. For corporations, 'worthy' projects are highly visible, foster a positive image that counteracts negative associations stemming from any environmentally-damaging practices they are engaged in, and promote the notion of a corporation as 'benefactor'. Corporate executives may feel they will get more extensive 'good news mileage' out of backing the 'cute and furries', the charismatic mega-fauna or the more spectacular habitats such as rainforests, than they will from the more obscure reptiles, invertebrates or small nocturnal species which may also face endangerment or extinction. Esso sponsors the Sumatran Tiger exhibit at Melbourne Zoo. Here is an example of a charismatic species that zoos have long favoured for display. The tiger is also consistent with Esso's brand identification. Moreover, the corporation's support for this highly endangered species might help to offset any negative images that might still be associated with the troubled environmental reputation of Exxon, Esso's parent company. The reputational effects of the Exxon Valdez oil spill have lasted many years, galvanising public attention about environmental disasters and leaving a negative opinion in many people's minds about Exxon's response to the crisis.[37]

There are other high profile corporate supporters of zoos, which are resource-extraction companies with controversial environmental reputations, such as Western Mining Corporation. What kind of messages is (knowingly or unknowingly) conveyed to the public when corporations consistently support certain kinds of zoo conservation projects or exhibit developments? How much does backing zoos help these companies develop a Good Samaritan profile that counteracts their more questionable stances towards environmental issues? Zoos may be in danger of espousing (however indirectly or unknowingly) the values of the business community that are incompatible with or that contradict ecological principles.

Sponsors of zoo exhibits always receive some kind of signage and/or brand identification in exchange for their support. The degree of visibility (the size of sign, sponsor's name and number of signs), style and wording of these signs will vary from one zoo to another. For example, BHP Steel supplied the steel for signs at the Melbourne Zoo. The corporate logo appears in relatively small print and is not included on all the steel signs in the zoo. The 'brand identification' in this case is subtler than other examples. Here BHP's provision of materials is less like advertising and more closely resembles philanthropy.

In contrast, McDonalds' sponsorship of zoos is readily visible and is a prerequisite for a sponsorship agreement. The corporation's regional offices currently sponsor several exhibits in Australasian zoos: the McDonald's South-East Asian Rainforest at Adelaide Zoo; the McDonald's Orang-utan Rainforest Home and the McDonald's Gorilla Forest at Taronga Zoo; the McDonald's Hippo Beach at Western Plains Zoo; and the McDonald's Rainforest at Auckland Zoo. The company's name is prominently featured in signs, banners and promotional materials for each of these exhibits.

Is this support for zoos' rainforest exhibits a function of an intentional, co-ordinated, corporate strategy specifically designed to counterbalance any negative views about McDonald's? Certainly the company is concerned about its reputation and market share. It spends approximately $2 billion a year on advertising. It has also had some significant public relations problems relating to environmental issues, such as its waste disposal practices[38] and the controversial libel suit it brought against two Greenpeace activists in England. As part of a broader campaign against the disreputable practices of multi-national corporations, the activists were distributing leaflets outside a London McDonald's restaurant. The brochures accused McDonald's of contributing to the destruction of South American rainforests by purchasing beef in areas that had been clear-felled, producing food which causes heart disease and cancer, and creating poor working conditions in its shops. The presiding judge on the case ruled that the defendants had been libellous because they had insufficient evidence to make these claims against McDonald's. Yet the judge also found the two Greenpeace members *had* shown that McDonald's 'exploits children' with its advertising, falsely advertises its food as nutritious, risks the health of its long-term regular customers, is 'culpably responsible' for cruelty to animals reared for its products, is 'strongly antipathetic' to unions, and pays its workers low wages.[39]

It may be too long a bow to draw to say that this damaging case against McDonald's can be directly linked to the corporation's support for zoo rainforest exhibits in Australasia. Sponsorship decisions within McDonald's are made at a regional level and often result from *zoo professionals* approaching McDonald's marketing staff in their respective regions—or States as is the case in Australia. Yet the striking symbolism and paradoxes in the corporation's support for zoos warrant mention, as they have the potential to compromise zoos' role in conveying sound and ecological educational messages to the general public.

McDonald's-sponsored zoo exhibits provide valuable vehicles for public relations and advertising platforms, but more insidiously the hidden message in these displays is that McDonald's is directly caring for the rainforest and its animal inhabitants. At the Adelaide Zoo, the rainforest exhibit is promoted by encouraging would-be visitors to 'see the first rainforest bred in captivity'. Leaflets encourage visitors to 'walk through and you'll soon discover an ever growing, ever changing environment that provides a natural habitat for many of South East Asia's endangered animals'. The Auckland Zoo's promotional brochure aligns the Zoo with McDonald's by asserting that both organisations' top priorities 'are children, families, fun, education and the environment'. The copy links several of the primate species on display to their wild, endangered counterparts. In addition to confusing concepts of 'natural' habitats and captivity, there are inherent contradictions in the message given to the public viewing these kinds of exhibits. That is, that saving the environment does not require substantial changes to individual, commercial and industrial activities; it merely calls for reconstructing a 'habitat' for rainforest species.

Western Mining Corporation also receives substantial signage at Healesville Sanctuary in exchange for its contribution to the Sanctuary's revegetation program. One sign appears in the

Sanctuary entrance where visitors pay their admission fees. The copy in these displays, 'a natural concern for Western Mining', suggests to the public that the company's support for the Sanctuary is equal to its broader environmental responsibility. The company does have an intricate environmental management system in place. Yet by aligning itself with the Sanctuary's progressive reputation in the zoo and wildlife conservation communities, Western Mining may also deflect any negative images of its less environmentally-friendly principles and practices. Interestingly enough, the company's managing director, Hugh Morgan, seems relatively unabashed about his negative feelings towards 'environmentalists', whom he attempts to lump into one homogeneous category:

> The road to power for ambitious revolutionaries is no longer the socialist road. But the environmentalist road, today, offers great opportunities for the ambitious, power-seeking revolutionary. Environmentalism offers perhaps even better opportunities for undermining private property than socialism.[40]

Hugh Morgan's views may provide an extreme example of hostile feelings towards environmentalists. Nonetheless, some corporate assistance in zoos could be interpreted as being somewhere between the rejectionist and accommodationist stance: an example of short-term, minimal responses by business to environmental problems, given an unwillingness or inability to move towards a greater understanding and acceptance of the long-range and critical nature of the issues.[41] It can be challenging to distinguish between corporations' notions of conscience and their range of corporate tools—such as public relations, government relations, competitive strategies, marketing orientation, legal compliance and issues management.

To what degree are zoos places where business can apply a salve to its conscience and avoid addressing some of the 'harder' environmental issues? Metropolitan zoos could be depicted as 'wildlife islands' in the midst of our cities, providing safe havens for endangered species. It may be far simpler to support the management of endangered species in *captivity* than to help restore the ecosystem that that species depends on, or address systemic environmental pollution stemming from a corporation's practices. Zoos are often criticised by animal rights and welfare lobbyists, but they still provide a relatively safe worthy cause for corporations to back. Good news stories can be produced, generating political mileage, corporate support and warm feelings amongst the public who want to know that species are being 'saved'. Corporations wishing to be (or be seen to be) community-minded and environmentally responsible may have found the perfect sponsorship beneficiary in zoos.

Such tactics could further discredit zoos' conservation reputation. We've seen that some people in the zoo community have been concerned by what they believe to be overly-commercialised zoo practices. Parts of the conservation community have shared these concerns (see Chapter 2). When interviewed, many of those people highlighted their distrust about zoos' 'money-making ventures' which they believed corrupt zoos' more munificent goals. Where there are misgivings about zoos amongst decision-makers in the conservation community, zoos may miss out on opportunities to participate in endangered species breeding programs.

Zoos also risk alienating visitors or members of the general community who are educated about the complexities of environmental politics or who have strong environmental sensibilities. For example, one visitor noted in his interview that, 'McDonald's is everywhere in this zoo . . . the human race [must be] in serious decline'. This may very well be an isolated example. Nonetheless, zoos have a responsibility to educate zoo visitors or members of the community who do not have extensive knowledge about environmental problems. There are several implicit messages to be concerned about when zoos sell corporations' sponsorship packages that provide extensive and favourable exposure for that corporation (and the zoo). The zoo is saying to the rest of the community that business is their partner and that business is a concerned and environmentally responsible citizen. Business can certainly be sincere about its support for zoos or for conservation. However, business's motivations for supporting environmental reform are typically complicated and hinge on a host of self-serving interests. And not all its responses are equally effective at restoring biological diversity. When zoos suggest otherwise, they limit their capacity to provide progressive environmental education.

Finally, there is a certain faulty logic to expecting zoos (or other cultural, educational, health, and environmental institutions) to seek funding from *private* sponsors to pay for the cost of fulfilling their *public* good obligations. Public good responsibilities are collective —they serve the interests of many. Conversely, private firms and individuals advance the interests of a few. When private firms or individuals claim to be good public citizens by virtue of their support of zoos for example, they are in effect distorting public good objectives. They are not necessarily qualified to judge what is needed to meet (and fund) public objectives. Rather, this was what governments, in close consultation with the community, were ideally set up to do.*

Conclusion

There seems to be little doubt what the proper 'business' of Noah's Ark was. Noah had a clear mission, the ultimate public good endorsed by the highest authority, and he commanded a vessel that was built strictly for that purpose. The biblical tale ends with Noah achieving his goal. In contrast, the modern Ark seems to be, in effect, dragging its anchor, caught up in ambiguity about how myriad economic efficiencies and the public good benefits of conserving biodiversity can be achieved.

The future of zoos' administrative and commercial apparatus seems more secure than what lies in wait for education, research and conservation activities in zoos. This course has been supported by certain social and political factors that have influenced decisions about how to obtain and distribute different kinds of resources to and within public organisations. Economic rationalism is one topical and highly persistent political fashion that has infiltrated corporate management frameworks of the public *and* private sectors. These managerial values have serious ramifications for all environ-

* I am grateful to Professor Max Neutze for these points.

mental administration in general and for zoo conservation policy in particular. Decision-making is predicated almost entirely on a highly ordered view of the world and on a series of different economic instruments to measure the effectiveness and efficiency of programs. This orientation typically empowers certain politicians, senior government officials, zoo board members and zoos' most senior managers. It also devalues programs (and their personnel) that do not provide any obvious financial benefits to zoos, and whose true worth and success may lie beyond any numerical measurement.

Such policy commitments could be shifted if all members of the zoo policy community are encouraged to openly and critically consider the full range of values that shape their policies and programs. The next chapter will consider some fora where this kind of debate has taken place and helped the zoo community achieve greater insights into its practices and potential for assisting biological conservation. However, further progress towards becoming the environmental resource centres endorsed by the World Zoo Conservation Strategy will require that the zoo community continue this discourse with greater knowledge of the full range of environmental values and the complexities of environmental policy-making and administration in Australia and overseas.

7
Sailing into Unknown Waters

And it came to pass in the six hundredth and first year . . . the waters were dried up from off the earth: and Noah removed the covering of the ark, and looked, and, behold, the face of the ground was dry . . . and Noah went forth, and . . . Every beast, every creeping thing, and every fowl, and whatsoever creepeth upon the earth, after their kinds, went forth out of the ark.

Genesis 6: 19, 20 (King James version)

The zoo has had considerable success reinventing itself, responding to changing social, economic, environmental and political contexts. Its evolution from private menagerie to public institution is perhaps the most significant shift. The early menageries served merely the interests of royalty and other élites. In contrast, as a public institution, the zoo is now beholden to serving a common interest. Most of today's zoos have several public good objectives, namely conservation, education, research and recreation.

Zoo professionals face numerous challenges as they try to fulfil such admirable goals. Finite financial, organisational and spatial resources, and flawed conservation methods limit the modern Ark's capacity to restore more than a small selection of species to the wild. Research programs have gone a considerable way to improving the quality of life for captive animals and integrating captive breeding programs with field conservation programs, but this preference for biological knowledge will not be sufficient for advancing the zoo's evolution to 'environmental resource centres'. Little research has been applied to understanding the broader setting of environmental problem-solving and how decisions and policies made by organisations, groups and individuals in the zoo community are related to that larger context. In addition, education programs are limited by an unresolved debate about the extent to which the zoo should confine itself to servicing recreational objectives and teaching people about biology and animals, or broaden its role to promoting critical thinking about a wide range of environmental issues.

There are certain structural and ideological factors that underlie these constraints. Rigid administrative rationality, which informs many organisational processes, actually favours a more conservative role for the zoo. Zoo professionals who want to incorporate more progressive ecological ideals into zoos' principles and programs must work against these arrangements rather than be supported by them. The zoo is also challenged by an economic rationality that has been dominating many problem-solving contexts throughout Western society for the last several decades. This political fashion empowers élites, seeks to privatise public goods and positions commercial values above ecological, social, health and educational imperatives. In the zoo, those who hold this world view downplay any conflict between economic and ecological values and claim that the zoo is doing all it possibly can—given of course what limited resources are available to it.

Broadening Ark Horizons

These contemporary policy struggles, as well as the Ark's historic evolution, arise from certain social and environmental values. At the beginning of this book, I alluded to the varied beliefs systems that include diverse definitions of and approaches to environmental problems. It is worth returning to the question of how values inform policy and to consider those values in more detail.

There are numerous ways to understand the values that inform different environmental discourses, those shared and contested stories that people use to interpret information about environmental issues.[1] Table 14 shows a spectrum of environmental values.[2] It has two primary themes which exist at either end of the spectrum; Technocentrism or technological environmentalism and Ecocentrism or ecological environmentalism. Technocentrism places ultimate faith in rational, scientific approaches that value non-human nature primarily on the basis of its *usefulness to people*. Advanced technologies and economic rationalism are used primarily to achieve *material* well-being. Conversely, ecocentrism challenges technocentric stances. Ecocentrism promotes the idea that non-human nature has an intrinsic value and is therefore worthy of our moral consideration. It also emphasises ecological limits to growth and low impact technologies.

Table 15 shows us that these values do not occur in mutually exclusive categories. They are part of a continuum. Both collectively and in specific organisations, not all zoo programs further the same environmental objectives or interests to the same degree, nor in a similar manner. In fact, zoos' principles and practices span the *entire* spectrum of environmental values, from the technocentric ideals manifest in economic rationalism, managerialism and high

Table 14 Environmental values and associated ideologies and methods

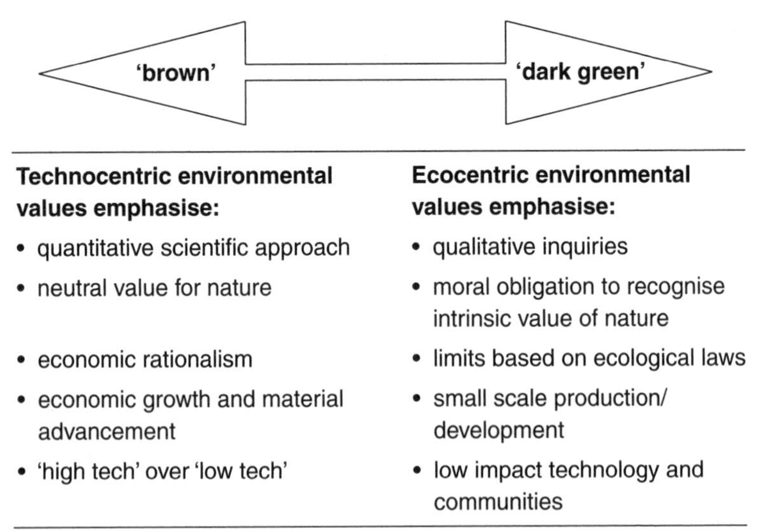

Technocentric environmental values emphasise:	Ecocentric environmental values emphasise:
• quantitative scientific approach	• qualitative inquiries
• neutral value for nature	• moral obligation to recognise intrinsic value of nature
• economic rationalism	• limits based on ecological laws
• economic growth and material advancement	• small scale production/ development
• 'high tech' over 'low tech'	• low impact technology and communities

SOURCE: Adapted from T. O'Riordan, *Environmentalism*, Pion, London, 1981.

Table 15 Environmental values informing zoos' principles and practices

'brown' ⟵		⟶ 'dark green'
Technocentric environmental values emphasise:		**Ecocentric environmental values emphasise:**
• exploitative: no restrictions on animal use	⟵ **attitudes to nature (animals)** ⟶	• animal rights/welfare primary factor in decision making
• commercial imperatives	⟵ **zoos' role** ⟶	• ecosystem conservation
• charismatic, exotic species in artificial conditions	⟵ **animal collection, maintenance and display** ⟶	• native flora & fauna from local ecosystems in natural groupings
• high technology, artificially-assisted reproduction	⟵ **breeding** ⟶	• unassisted natural breeding of naturally-occuring populations
• basic animal biology & taxonomy	⟵ **education** ⟶	• ecologically-oriented interdisciplinary environmental education
• quantitative inquiries in biological sciences	⟵ **research** ⟶	• quantitative, biological studies integrated w/qualitative social research
• autocratic leadership, strict hierarchies, corporatised management systems	⟵ **organisational practices** ⟶	• small organisations, highly-participative decision-making, integrated responsibilities

SOURCE: Adapted from T. O'Riordan, *Environmentalism*, Pion, London, 1981.

tech scientific agendas to the ecocentric beliefs underlying ecosystem conservation and exhibitry, qualitative, interdisciplinary environmental research, environmental education and participative policy-making fora. Some zoo professionals favour elements from only the extreme ends of the spectrum. Still others sit somewhere in the middle, promoting values from both categories. Irrespective of their value stance, the choices made by zoo professionals'—and others involved in zoo policy-making—will be affected by particular issues and/or by conditions that are specific to their organisations and surroundings. Eventually, these standpoints are reconciled in some way when contradictions or opposing sets of belief and action clash in zoos' decision-making systems. The outcome of these contests in zoos tends to resemble the broader trend in Western society: technocentric values are predominant in conventional and official environmental problem-solving arenas.

This situation is highly problematic for those in the zoo community seeking to stem the loss of biodiversity conservation and to teach people to care for their environment. Some suggest that the pre-eminence of technocentric ideals may be helping to create—not an enlightened society—but an 'unconscious civilization' that fools itself by continually believing in the power and worth of economic and managerial rationality, while neglecting the social and environmental costs of doing so.[3] A growing body of work suggests that we have failed to create a more socially just and ecologically sustainable society (and zoos?) to date, because of a general unwillingness to recognise the mistaken assumptions underlying technocentric values:

- that are no limits to economic growth (even though our economic systems are embedded in finite ecosystems);
- that 'progress' is equal to ever higher standards of living and greater material wealth;
- that any problem or human behaviour can be readily calculated and managed accordingly;
- that greater efficiencies can be gained by ever-increasing amounts of organising and control.[4]

Alternatively, more reflective and humanitarian societies, institutions and organisations will be created when policy decisions are informed by principles of equity and social and environmental justice.[5]

Defending Conventional Practice

The zoo will need to consider its capacity for such reflective practice in order to improve its environmental performance and provide more support for biodiversity conservation. As zoo professionals have integrated conservation, education and research programs into the zoo's programming over the last few decades, they have highlighted that the institution has certainly made notable changes. Despite these achievements, 'the zoo' is often still viewed and treated more as a place of recreation and entertainment than as an institution renowned for its scholarly, scientific or conservation achievements. These doubts about zoos occur in the wider community and inside the zoo community itself. Most recently, we have also seen that some of zoos' contemporary practices still embody principles from previous centuries and embrace technocentric environmental thought. These tendencies suggest that, instead of a fundamental shift in policy, the zoo's new identity is rather more tenuous. Zoo professionals may have become more adept at meeting their problems with rationalisations that merely convey the *impression* they know what they are doing in order to impress others, to convince themselves that all is well and to show that they have the ability to cope.

Part of the problem may lie in the fact that in responding to criticisms and the need for change, sections of the zoo community sometimes react with defensiveness, which inhibits their ability to question their own standpoints and to further advance the zoo's evolution. For example, the zoo community's historical allegiance to the mainstream scientific community has lent some credibility to its endeavours, such as high tech *ex-situ* research. However, this close association does not seem to have strengthened the zoo community's enthusiasm or responsibility for participating in debates about the social responsibilities of science and the ways that scientific knowledge and technology have failed to solve society's problems. Another example is the considerable time and resources that are spent promoting official policies and position statements. Such actions can entrench conventional practice and misrepresent or overstate zoos' ability to deliver tangible and substantial benefits to wildlife and people. Like any defensive reaction, these policies may protect individuals, groups or organisations from experiencing embarrassment or threat, but they also prevent them from identifying and reducing the *causes* of that embarrassment or threat and from understanding the underlying values of their own positions.[6]

Some zoo professionals are painfully aware that these conservative stances inhibit more reflective practice and limit the zoo's full potential. Such individual insights are very important. They are not sufficient for achieving more substantial change, which would require critical thinking or 'learning' at an *organisational* and *institutional* level. Where there are discrepancies between zoos' policy rhetoric and their program realities, a substantial proportion of zoos' evolution will have resulted from simple (or single-loop) learning.[7] That is, while zoo professionals have helped the zoo to evolve by adjusting some programs to reflect shifting political, social and environmental concerns, many basic operating principles remain unchanged. Greater change and improved performance in zoos would come from more complex (or double-loop) learning.[8] Organisations that exhibit this level of learning encourage and value openness and reflectivity and accept error and uncertainty as inevitable features of life in complex and changing environments. Essentially, members of the zoo community would have to observe and correct their mistakes to such an extent that they ask not just 'How well are we doing?', but also 'Does it make sense to be doing it?',[9] especially in relation to the larger context of environmental conservation?

This kind of learning is certainly not being encouraged where economic, administrative and technical rationality is allowed to have a significant influence on the policies and programs of zoos (and other institutions). Rather, complex learning is often frustrated by these forces, which impose rigid principles and programs that make it very difficult to gain new knowledge or change work routines. Furthermore, corporatised management practices and bureaucracies' fundamental organising principles block the learning process through their low tolerance for uncertainty that both encourages and rewards defensive behaviours. Under these conditions, people are pressured to attend to symptoms of problems—short term predicaments that can be resolved quickly with obvious, simple solutions—rather than causes. In so doing, they overlook or dismiss the more complex and fundamental problems that might be plaguing the organisation and the people working in it.[10]

A More Reflective Ark

Given the worsening environmental conditions and myriad social problems confronting modern society, there is now a stronger imperative than ever for the zoo's continued evolution. There are many zoo professionals who, individually and collectively, are making a concerted effort to turn their workplaces into learning organisations and to ensure that the zoo's transformation is anything but conventional. The common thread linking these disparate efforts is a willingness to deeply question what society has come to know as 'the zoo'.

The two conferences mentioned earlier are also worth revisiting, because they made important contributions to zoo policy during the early and mid-1990s. Zoo professionals, activists, conservation biologists and philosophers helped to maintain the forward momentum of zoo evolution when they attended a symposium, held in Atlanta, Georgia in 1992. At this forum, the participants' vigorously debated the future of zoos and aquaria, the treatment of animals in captivity and questioned whether the individual, the species or the ecosystem is the most important focus in conservation efforts. Their deliberations were eventually published in the book *Ethics on the Ark*. In the foreword to this book, a poignant call was made for the zoo community to embrace critical thinking and complex learning by recognising the value in varied standpoints about zoos—even the unfavourable ones. It said that, 'The only way for self-examination to work is if it is completely honest . . . uncompromising honesty is the most striking feature'.[11]

The form and content of the 1996 ARAZPA/ASZK annual conference hosted by Healesville Sanctuary in Victoria were also significant for zoo reform. An important factor in their eventual success was the open and participative strategy used for organising the event. The Sanctuary's Director trusted his staff's advice and gave them relatively free rein to do as they saw fit. Members of the organising committee wanted to engage in meaningful and stimulating debate about how the modern zoo was evolving. They were frustrated by conventional approaches to conferences and were seeking an innovative focus and format that would produce constructive outcomes. The topics they selected were designed to help the zoo community look critically at itself, questioning the sincerity and effectiveness of current policies and programs. Seeking to overcome the tyranny of distance and capitalise on having so many of their interstate and international colleagues in the same place at the same time, formal papers were kept to a minimum, while numerous workshops were built into the schedule. This provided a setting where zoo staff from varied places and organisational levels could discuss issues that concerned them all in an informal atmosphere which would minimise anyone's fear of reprisal or recrimination.

There was a strong emphasis on openly acknowledging differences of opinion as a way to discover possible future directions for the zoo. Greg Hunt, conference mediator, said that he felt that the conference was highly successful 'if not in reaching consensus on these issues (and how could that be possible?), but in clarifying the issues that we should consider in our respective organisations':

> For this success we must thank the preparedness of all participants to speak openly and fearlessly, to enter into the spirit of debate and to accord respect to opposing views. That such debate is flourishing within our membership is one of the most encouraging signs that points to an assured future for zoos.[12]

Hunt is right to point out that total consensus on the Ark's ultimate course may never be achieved, especially since conflicting values are an inherent feature of policy landscapes such as wildlife conservation. Nonetheless, the American and Australian events provide examples of open, participative, constructive and proactive self-examination that—if continued—would certainly help to clarify where the Ark is sailing to.

Peter Stroud, a curator at the Melbourne Zoo, has a vision of the ideal zoo that is drawn partly on the visions of many others. He believes that the zoo of tomorrow should not be confined to the traditional identity of a recreational institution, which services the interests of a homogeneous audience. Stroud's zoo is a place that positions conservation, education and research as things that the public values, and so it has different kinds of programs for different kinds of audiences. At the centre of the zoo would be exhibits based on different habitats across the world—focused on providing visitors with the story of 'life on earth'. The outer ring of the zoo would feature the Zoo's different *ex-situ* and *in-situ* conservation and research programs. Interested visitors could take 'behind the scenes' tours, participate in a variety of education programs, and use interactive multi-media displays to view national and international sites outside the zoo where actual conservation and research programs are taking place. Visitors could even see these sites through special tour packages. His vision also incorporates a more holistic organisation, one where zoo staff would be organised into teams servicing the different exhibit and program areas and where decisions were made in highly participative fora.

Graham Mitchell, former Director of the Melbourne Zoo, believes that zoos could make more meaningful contributions to biodiversity conservation by having closer relationships with government and non-government conservation agencies, possibly even becoming shopfronts for them. These organisations—at state and national levels in Australia—continually struggle to find sufficient resources, especially for providing public education. Professor Mitchell sees a vast potential for the mutual exchange of resources, values, skills and knowledge between these organisations and zoos. There is a wealth of expertise in environmental education, marketing, small population management and captive animal husbandry in zoos. Staff in government and non government agencies have extensive knowledge and skills in broader concepts of ecosystem conservation and environmental management. By working in a co-ordinated fashion the various agencies ideally would fill voids they have previously not been able to. Zoos would be able to widen their brief beyond wildlife conservation and other agencies would gain further access to an interested public. In some ways, the wildlife parks in Darwin and Alice Springs, which are managed by the Northern Territory's Parks and Wildlife Commission, provide the closest real example of this vision.

Some zoo professionals' ideal of a new zoo essentially targets the dilemma inherent in the zoo being caught between being, on one hand, a non-profit organisation and, on the other, a commercial entity. The zoo falls short of doing justice to either, because there is insufficient appreciation of the differences between the two different types of organisations. One zoo professional suggested that Western society should totally reject the idea that zoos lead the way in the 'recreation business'. Instead, zoos would function like museums, art galleries or national parks, which charge nominal or no admission charges and have more substantial government support. This person's vision certainly provides some food for thought. If the zoo was relieved from the pressure to increase its commercial capacities, perhaps it would be more accountable to the imperative to provide a public good, and more resources could (ideally) be devoted to conservation, education and research imperatives. Marketing and management strategies could then be revised to be more sensitive to the unique needs of non-profit organisations. Since these entities have non-financial goals, success and failure could no longer be measured in terms of financial indicators. The ultimate aim of marketing would be to ensure that revenues were channelled back into conservation, education and research programs, and that none of these programs' integrity would be threatened by excessive commercialisation. Rather than seeing other cultural venues such as museums, art galleries and aquariums as competition, those marketing the zoo could join forces with these institutions. Selling discount packages that provide entry to all or several of those venues could foster greater public support and visitation.

All Hands—Abandon Ship!?

The visions of a future zoo discussed above suggest that 'the Ark' might be an outdated vessel, the metaphor no longer be suitable for depicting what the zoo can or should be. Ideas about institutions and organisations are always based on implicit messages or metaphors that persuade us to see, understand and manage situations in a particular way. The strength of metaphors comes from providing insights that might not have been considered before. Metaphors also have weaknesses and can distort our thinking. That is, in creating ways of seeing, metaphors also create ways of *not seeing*.[13]

The Ark metaphor has been important for the zoo's evolution. During the middle of the twentieth century, it provided a fresh vision for the institution. Zoo professionals and the general community could imagine the zoo, not as a consumer of nature, but as a (temporary) safe haven where wildlife could ride out the storm of environmental degradation. The central cultural themes of collecting and collections have also been relevant to the Ark mythology. They are desire and nostalgia, saving and loss and the urge to erect a permanent and complete system against the ravages of time.[14] The urge to collect remains an important part of human culture and signals our amusement with and longing for worlds that existed before their contents were collected. Noah might well be considered as the supreme collector. He acted quickly, decisively, thoroughly and overlooked nothing. Noah's collection was a 'unique bastion against the deluge of time . . . and Noah, perhaps alone of all collectors, achieved the complete set'.[15] He was an extreme example of a collector, one driven by a higher cause. Noah was rescuing his collection from extinction, and that collection was in effect equal to the world's salvation.

Noah was also victorious because he knew exactly what his mission was, which in turn enabled him to build a vessel specific to that purpose. He also had sufficient resources to save a representative sample of every living thing on earth. Finally, while he had to endure a long passage at sea, the journey was finite. Noah was able to guide the Ark to its final destination—he reached landfall after the floodwaters receded.

Herein lie some essential differences between the biblical and modern Arks, and the reasons why the Ark metaphor may have outlived its usefulness to modern zoo professionals and the broader community. Zoo professionals of today face innumerable ambiguities that the Noah of yesteryear did not struggle with. Today, it seems as if the zoo community is not sure what destination it is guiding the modern Ark towards. The zoo community also seems distracted by the complexities of the journey, which in turn cloud the horizon. Perhaps in the process of keeping all the vessels of the modern Ark fleet afloat and sufficiently and appropriately managed, resourced and co-ordinated, members of the zoo community occasionally lose sight of which 'final destination' they are heading towards.

Moreover, trying to spot 'landfall' while the modern Ark floats in a sea of social, political, economic and ecological complexities cannot be an easy task. Zoo professionals have to cope with the fact that 'landfall' is not likely to look anything like it did when the biblical Ark was set ashore. Modern environmental knowledge and management have advanced significantly; however, there is plenty of evidence to suggest that the 'flood waters' are not likely to recede any time in the near future, and that they have inflicted extensive and permanent ecological damage in the interim. Here again, the culture of collecting presents a dilemma, not just for zoo professionals but for all those trying to preserve species and habitats alike. Modern conservation efforts embody nostalgia for worlds that are rapidly disappearing and are not likely ever to return. In effect, ecological targets keep shifting, as decisions about what to 'save' and how to save it must be continually modified. As I stated earlier, zoo professionals and the general community have yet to agree about what the role of the zoo in relation to environmental conservation can or should be.

This does not mean that members of the zoo community have not thought about or do not continue to think about what kind of institution they want the zoo to be. George Rabb's vision of zoos as environmental resource centres is one vision that could be used to replace the Ark. I am hopeful that this book has and will continue to complement these efforts at imagining a new and promising future for zoos by reiterating some of the persistent concerns about zoos. I have also attempted to open up alternative pathways for thinking about the problems and opportunities zoos face, namely that the varied perspectives about appropriate roles for the zoo in conservation are essentially a subset of broader environmental discourses, which include:

- questions about the limits to the Earth's natural resources and what threat their use poses to the Earth's capacity to support human and non-human life;
- whether we can adequately address ecological problems by relying on existing administrative, technical, economic and scientific frameworks; and
- how well sustainable development is delivering its promise to find a balance between economic and ecological values.[16]

If zoos around the world are to evolve into 'environmental resource centres' that have significant and favourable impacts on biodiversity conservation and education, zoo professionals and the general community need to understand how the zoo's policies and actions are related to these major discourses. Yet understanding the zoo in the context of broader ecological problems and environmental policy contexts necessitates that the zoo community be willing and able to openly and honestly critique—not just the Ark metaphor—but the values and assumptions that are embodied in that image. No single theory or metaphor will provide a satisfactory vision for the zoo's future. Rather, multiple imaginings of the zoo's evolution are needed. Whatever choices are made, the zoo's evolution will have advanced in the 'right' direction when all views—the affirming *and* critical stances alike—find an *equal* voice in zoos' short- and long-term decision-making and when zoo professionals devote sufficient attention to how their social, political and organisational settings shape the identity of their institution.

Appendix 1: Research Programs in Australian Zoos

In Victoria, the Zoological Parks and Gardens Board's formal research projects for 1998 show a bias towards mammalian species. Sixty-four per cent of its thirty-three projects focused on mammals (with marsupials and monotremes accounting for half of the total projects), 24 per cent looked at birds, and 6 per cent were on invertebrates. These projects were spread fairly evenly across different disciplines in the biological sciences. Five behavioural studies looked at four different mammalian taxa (primate, hoofstock, marsupial, monotreme—the platypus) and one bird species. The eight reproductive projects also inclined towards mammalian species: four were general studies of marsupials (with one looking at a particular species of wallaby), two projects examined platypus, one was on seals, and one was oriented towards endangered species in general. The six taxonomic/genetics projects included studies of two marsupial species, two bird species, one invertebrate species, and one general study of zoo animals. The six examinations of animal health comprised studies of one species of bird, one of birds of prey, and four of marsupial species (two of those on the same species). The six ecological research projects included work on three marsupial species, three bird species (for two projects), and one invertebrate species.

The Board's publications showed a more even coverage of taxa and subject matter. Twenty-four per cent of publications focused on mammals, 15 per cent on birds, 13 per cent on invertebrates, 4 per cent on reptiles, and 8 per cent on plants. The remaining articles were about education (19 per cent) and zoo conservation and management (17 per cent) issues. Most of the pieces on zoos' conservation role appeared in newsletters.

The largest establishment, the Zoological Parks Board of NSW, maintains a prolific rate of publication. Seventy-seven publications were listed in the Board's annual report for 1997, most of those pieces appearing either in journals (a mixture of scientific and non-scientific) or conference proceedings. Over half those pieces had mammals as their primary focus. Zoo management and zoos' role in conservation accounted for another 19 per cent of the articles. Essays on birds (12 per cent), reptiles/amphibians (3 per cent), and fish (3 per cent) made up the remainder of publications. Animal health/disease (19 per cent), behaviour (15 per cent), and physiology (12 per cent) most heavily represented the scientific disciplines of these articles. There were similar biases for the Board's research projects. Studies of mammals accounted for over half the total, with eleven of those projects focusing on marsupials and another five on monotremes (the platypus, primarily). Research on birds (17 per cent), reptiles/amphibians (11 per cent), zoo management matters (8 per cent) and zoo animals in general (6 per cent) made up the remainder of activities. The disciplinary orientation of these studies was distributed primarily among the fields of behaviour (23 per cent), reproduction (19 per cent), genetics (17 per cent), physiology (15 per cent), and husbandry (15 per cent). The remaining 26 per cent included taxonomic, disease, ecology, and miscellaneous research.

The Royal Zoological Society of South Australia produced twenty-one articles in journals, conference proceedings, reports and newsletters. Eight of those pieces were about mammals, all but one of them focusing on marsupials. Another five articles were on birds and two were on education. Zoo management and conservation management was discussed in five conference papers. Sixteen research projects were listed in the Society's annual report for 1998. Virtually all of those projects (twelve) were about mammals; eight of those twelve examined marsupial species. Interestingly, two projects examined plants and another two were studies of environmental contaminants in the zoo setting. Smaller arrays of scientific disciplines were represented in the Society's projects: four behavioural studies, two examinations of behavioural ecology; four reproductive studies; and two animal health inquiries.

The Perth Zoo had comparable leanings in its research projects. Of the projects listed as completed and ongoing in its 1998 annual report, 51 per cent were studies of animal behaviour. Eighteen per cent and 14 per cent of the projects, respectively, examined animal reproduction and genetics. The remaining 17 per cent of the projects looked at physiological, husbandry, animal health/disease, and demography issues. The bent of these inquiries towards particular taxa was the most exaggerated at Perth Zoo. Some single species receive extended attention, reflecting the availability of specimens represented in the Zoo's collection and the specialised training and interests of the researchers. Six per cent of the research projects focused on birds, 4 per cent looked at reptiles, and 6 per cent highlighted plants or invertebrates. Eighty-two per cent of the research projects listed in the annual report focused on mammals. That body of work included studies of marsupials (31 per cent), primates (28 per cent, of which 55 per cent were on the Orang-utan), hoofstock (13 per cent, of which 80 per cent were on the giraffe), and a mixture of species (28 per cent), such as carnivores, meerkats, and otters.

Appendix 2: International and National Strategies Relevant to the Role of Zoos in Conservation

Convention on Biological Diversity (1992)	Article 13. Public Education & Awareness The contracting parties shall (a) Promote and encourage understanding of the importance of, and the measures required for, the conservation of biological diversity, as well as its propagation through media, and in the inclusion of these topics in educational programs; and (b) Cooperate, as appropriate, with other States and international organisations in developing educational and public awareness programs, with respect to conservation and sustainable use of biological diversity.
The World Zoo Conservation Strategy (1993)	The overall aim is to help conserve Earth's fast-disappearing wildlife and biodiversity . . . Zoos should support the objectives of the World Zoo Conservation Strategy (and related documents) by: (3) promoting an increase of public and political awareness of the necessity for conservation, natural resource sustainability, and the creation of a new equilibrium between people and nature.
National Strategy for Ecologically Sustainable Development (1992)	Embracing ESD ultimately rests on the ability of all Australians to contribute individually through modifying everyday behaviour, and through opportunities open to us to influence community practices . . . Challenge: To ensure that the implementation and further development of this Strategy benefits from informed community participation, and that progress towards ecologically sustainable development is supported by community understanding and action.

	Objective 32.1 To develop a high level of community awareness and understanding of the goal, objectives and principles of this ESD Strategy
	Challenge: To increase awareness and application of ESD principles and approach in education and training policy and programs. Objective 26.1 to incorporate ESD principles and approaches into the curriculum, assessment and teaching programs of schools and higher education Objective 26.2 to develop and improve vocational education and training programs which incorporate ESD principles and which will give practical skills in achieving ESD
National Strategy for the Conservation of Biological Diversity (1992)	Objective: increase public awareness of & involvement in conservation of biodiversity . . . accessible personal action guides explaining ways individuals & groups can help
National Strategy for the Conservation of Australian Species and Communities Threatened with Extinction (1992)	Objective 1: To develop and implement a national education program on endangered and vulnerable species and ecological communities directed towards all sectors of Australian society; Objective 11: To promote and participate in the international efforts directed to the conservation of endangered and vulnerable species and ecological communities; 11.7: Encouraging zoos, botanic gardens and other living museums to develop conservation priorities at an international and regional level and then coordinate management of their collections to reflect those priorities.

SOURCES: Australian and New Zealand Environment and Conservation Council, *National Strategy for the Conservation of Biodiversity*, Department of Environment, Sports and Territories, Canberra, ACT, 1992; The Captive Breeding Specialist Group and the World Zoo Organisation, *The World Zoo Conservation Strategy*, Chicago Zoological Society, Chicago, IL, 1993; Ecologically Sustainable Development Committee, *National Strategy for Ecologically Sustainable Development*, Australian Government Publishing Service, Canberra, 1992; Endangered Species Advisory Committee, *National Strategy for Conservation of Australian Species and Communities Threatened with Extinction*, Australian National Parks and Wildlife Service, Canberra, 1992; United Nations Environment Program, *Convention on Biological Diversity*, 1992 www.biodiv.org/chm/conv/art0.htm

Notes

Introduction

[1] S. Hastings, *Noah's Ark and Other Bible Stories*, RD Press, Surry Hills, NSW, 1996, p. 12.
[2] Mullan, *Zoo Culture*, p. xvii.
[3] Captive Breeding Specialist Group, *The World Zoo Conservation Strategy*, p. x.
[4] Rabb, 'The Changing Roles of Zoological Parks in Conserving Biological Diversity', p. 162.
[5] Clark, *Endangered Species Recovery*.

1 Assembling the Ark

[1] M. Twain, *Letters from Earth*, Harper & Row, New York, 1962, p. 21.
[2] Singer, *In Defense of Animals*; Ryder, *Animal Revolution*.
[3] Ryder, *Animal Revolution*; Singer, *Animal Liberation*.
[4] Glacken, *Traces on the Rhodian Shore: Nature and Culture in Western Thought from Ancient Times to the End of the 18th Century*, pp. 173–5.
[5] Ryder, *Animal Revolution*.
[6] Thomas, *Man and the Natural World: Changing Attitudes in England, 1500–1800*.
[7] Cherfas, *Zoo 2000*, p. 16.
[8] Glacken, *Traces on the Rhodian Shore: Nature and Culture in Western Thought from Ancient Times to the End of the 18th Century*.
[9] P. B. Ebrey, *The Cambridge Illustrated History of China*, Cambridge University Press, Cambridge, 1996, pp. 209–12.
[10] Mullan, *Zoo Culture*, pp. 103–4.
[11] Glacken, *Traces on the Rhodian Shore*.
[12] Cherfas, *Zoo 2000*, p. 22.
[13] See Mullan, *Zoo Culture*, pp. 89–112.
[14] Ibid., p. 97.
[15] Ryder, *Animal Revolution*; Thomas, *Man and the Natural World*.
[16] Mullan, *Zoo Culture*, pp. 101–3.
[17] Ritvo, *The Animal Estate*, p. 207.
[18] Cherfas, *Zoo 2000*, p. 35; Mullan, *Zoo Culture*, p. 109.

[19] J. Bierlein, 'Interpretation and the Zoo Renaissance', *Legacy*, vol. 2, no. 2, 1991, p. 13.
[20] M. Brambell, 'The Evolution of the Modern Zoo', *International Zoo News*, vol. 40, no. 7, pp. 27–9.
[21] K. Anderson, 'Animals, Science, and Spectacle in the City', in J. Wolch & J. Emel (eds), *Animal Geographies: Place, Politics, and Identity in the Nature-Culture Borderlands*, Verson, London, 1998, p. 36; Mullan, *Zoo Culture*, pp. 110–14.
[22] T. Griffiths, *Hunters and Collectors: The Antiquarian Imagination in Australia*, Cambridge University Press, Cambridge, 1996, p. 12.
[23] Griffiths, *Hunters and Collectors*, pp. 12–16; N. Low, *Feral Future*, Penguin Books, Ringwood, Victoria, 1999, pp. 33–4, 195.
[24] de Courcy, *The Zoo Story*, p. 17.
[25] Jenkins, *The Noah's Ark Syndrome*, pp. 108–9.
[26] Ryder, *Animal Revolution*.
[27] Bostock, *Zoos and Animal Rights*, pp. 35–6; Ryder, *Animal Revolution*.
[28] Bostock, *Zoos and Animal Rights*, p. 29; Ritvo, *The Animal Estate*, p. 223.
[29] Mullan, *Zoo Culture*, p. 48.
[30] Zuckerman, *Great Zoos of the World*, p. 19
[31] See for example: Bostock, *Zoos and Animal Rights*, pp. 30–1.
[32] K. Frawley, 'Evolving Visions: Environmental Management and Nature Conservation in Australia', in S. Dovers (ed.), *Australian Environmental History*, Oxford University Press, Melbourne, 1994, pp. 216–34; J. M. Powell, 'Protracted Reconciliation: Society and the Environment', in R. MacLeod (ed.), *The Commonwealth of Science: ANZAAS and the Scientific Enterprise in Australasia 1888–1988*, Oxford University Press, Melbourne, 1988, p. 249.
[33] Low, *Feral Future*, p. 27.
[34] de Courcy, *The Zoo Story*, p. 132.
[35] T. Crosby, 'The Early Years: Evolution of the Zoo', in A. Meyer (ed.), *A Zoo for All Seasons: The Smithsonian Animal World*, Smithsonian Exposition Books, Washington DC, 1979, p. 30; G. Woodruffe, *Wildlife Conservation and the Modern Zoo*, Saiga Publishing Co. Ltd, Surrey, 1981, p. 4.
[36] Crosby, 'The Early Years', pp. 32–3.
[37] Frawley, 'Evolving Visions', pp. 216–34.
[38] Ryder, *Animal Revolution*.
[39] M. Greene, 'No Rooms, Jungle Vu', *The Atlantic Monthly*, December, 1987, pp. 62–78.
[40] Bierlein, 'Interpretation and the Zoo Renaissance', p. 12.
[41] Kellert, 'The Educational Potential of the Zoo and its Visitors', p. 11; Mitchell, 'Conserving Biological Diversity: a View from the Zoo', p. 12.
[42] J. Hatley and J. King, 'The Birth and Growth of the IZE', *Journal of the International Association of Zoo Educators*, no. 29, 1993, p. 206.
[43] R. Baker and G. G. George, 'Species Management Programs in Australia and New Zealand', *International Zoo Yearbook*, vol. 27, 1988, pp. 19–26.
[44] Quoted in Zuckerman, *Great Zoos of the World*, p. i.
[45] N. Duplaix and L. Grady, 'Is the International Trade Convention For or Against Wildlife?', *International Zoo Yearbook*, vol. 20, 1980, pp. 171–6.
[46] T. J. Foose and J. D. Ballou, 'Management of Small Populations', *International Zoo Yearbook*, vol. 27, 1988, p. 27.

[47] R. Garner, 'Wildlife Conservation and the Moral Status of Animals', *Environmental Politics*, vol. 3, no. 1, 1994, pp. 114–29; Ryder, *Animal Revolution.*
[48] Ryder, *Animal Revolution.*
[49] G. Mitchell, 'A Perspective on Zoos in a Changing Environment', *Australian Academy of Technological Sciences and Engineering—Focus*, no. 81, March/April, 1994, p. 25.
[50] Rabb, 'The Changing Role', p. 162.

2 Passenger Lists and Procedures

[1] United Nations Environment Program, *The International Convention on Biological Diversity*, 5 June, 1992, p. 3
[2] Primack, *Essentials of Conservation Biology*, p. 405.
[3] Ibid., pp. 405–7.
[4] Captive Breeding Specialist Group, *The World Zoo Conservation Strategy*, p. 49.
[5] Meeks, *Beyond the Ark.*
[6] Ibid.; Minta and Kareiva, 'A Conservation Science Perspective: Conceptual and Experimental Improvements', p. 277; Vrijenhoek, 'Natural Processes, Individuals, and Units of Conservation', pp. 74–5; McIntyre, 'Species Triage—Seeing beyond Wounded Rhinos', pp. 604–6.
[7] La Roe, 'Implementation of an Ecosystem Approach to Endangered Species Conservation', pp. 3–6.
[8] Clark, *Endangered Species Recovery.*
[9] Fiedler *et al.*, 'The Contemporary Paradigm in Ecology', p. 11.
[10] Groombridge, *Global Biodiversity*, pp. 569–71.
[11] Beck, 'Reintroduction, Zoos, Conservation, and Animal Welfare', p. 156.
[12] Snyder *et al.*, 'Limitations of Captive Breeding', p. 338.
[13] Ibid., p. 343.
[14] Ormrod, 'Showboat as Ark', p. 42.
[15] Zoological Parks Board of New South Wales, *Annual Report* 1992/1993, p. 75.
[16] World Society for the Protection of Wild Animals, *The Zoo Inquiry*, p. 30.
[17] Groombridge, *Global Biodiversity*, p. 569.
[18] Fiedler *et al.*, 'The Contemporary Paradigm in Ecology', p. 11.
[19] Woodruff, 'The Problems of Conserving Genes and Species', p. 77.
[20] Groombridge, *Global Biodiversity*, p. 569.
[21] Fiedler *et al.*, 'The Contemporary Paradigm in Ecology', p. 11.
[22] Captive Breeding Specialist Group, *World Zoo Conservation Strategy*, p. 74.
[23] World Zoo Organisation, *Zoo Futures 2005.*
[24] Australasian Species Management Program, *Regional Census & Plan*, p. 18.
[25] Rabinowitz, 'Helping a Species Go Extinct', pp. 482–8.
[26] Doyle and Kellow, *Environmental Politics and Policy-Making in Australia*, p. 145.
[27] Holland *et al.*, *Federalism and the Environment: Environmental Policy Making in Australia, Canada, & the United States*, p. 2.
[28] Dixon, 'Protection of Endangered Species'; Bates, *Environmental Law in Australia.*

[29] Hunter, 'Overview of Biodiversity (Species and Protected Areas) Provisions', pp. 41–4; Kennedy, 'The EPBC Act & Biodiversity Protection: A Conservation Organisation Perspective', pp. 45–55.

[30] Australasian Species Management Program, *Regional Census and Plan*, 1998, pp. 150–3; Australasian Species Management Program, *Regional Census and Plan*, 1994, p. 92.

[31] Australian Nature Conservation Agency, *Recovery Plan and Funding Proposal Guidelines*, Endangered Species Unit, Australian Nature Conservation Agency, Canberra, 1994, p. 7.

[32] For more detail see: Mazur, Contextualising the Role of Zoos in Conservation, pp. 312–17.

[33] C. Banks, 'The Striped Legless Lizard Working Group—An Inter agency Initiative to Save an Endangered Reptile', in *Saving Wildlife: Proceedings of the ARAZPA/ASZK Annual Conference*, 4–10 April, Currumbin Sanctuary, Queensland, 1992, p. 47.

[34] M. McCarthy, 'Population Viability Analysis of the Helmeted Honeyeater: Risk Assessment of Captive Management and Reintroduction', in Serena, *Reintroduction Biology of Australian and New Zealand Fauna*, p. 21.

[35] I. Smales, P. Menkhorst and G. Horrocks, 'The Helmeted Honeyeater Recovery Program: A View of its Organisation and Operation', in Bennett *et al.*, *People and Nature Conservation*, p. 35.

3 Research for Sailing or Docking?

[1] Barnes, *About Science*, p. 11.

[2] K. Benirschke, 'The Need for Multidisciplinary Research Units in the Zoo', in Kleiman, *Wild Mammals in Captivity*, p. 538.

[3] See for example: O. A. Ryder and A. T. C. Feistner, 'Research in Zoos: a Growth Area for Conservation', *Biodiversity and Conservation*, vol. 4, p. 671; M. Hutchins and W. G. Conway, 'Beyond Noah's Ark: the Evolving Role of Modern Zoological Parks and Aquariums in Field Conservation', *International Zoo Yearbook*, vol. 34, 1995, pp. 117–30.

[4] Jamison, 'Western Science in Perspective', pp. 131–44.

[5] Maguire, *Doing Participatory Research*, p. 11.

[6] L. Birke, *Women, Feminism and Biology: The Feminist Challenge*, Harvester Press, 1986, pp. 76–82.

[7] D. D. Murphy, 'Conservation Biology and the Scientific Method', *Conservation Biology*, vol. 4, pp. 203–4.

[8] Jamison, 'Western Science in Perspective ', pp. 131–40

[9] Ibid.

[10] Brunner, 'Science and Social Responsibility', p. 295.

[11] M. Cameron, Are Positivist Methodologies the Only or Even the Most Appropriate Methodologies on Which to Base Research in Environmental Studies?, Course materials for Environmental Politics, Philosophy and Ethics, Mawson Graduate Centre for Environmental Studies, University of Adelaide, Adelaide, 1996, p. 3.

[12] Graham, *Between Science and Values*, p. 291.

[13] Brunner, 'Science and Social Responsibility', p. 295.
[14] C. S. Holling, 'What Barriers, What Bridges?', in Gunderson *et al.*, *Barriers and Bridges*, pp. 3–36.
[15] H. Ross, 'Social R&D for Sustainable Natural Resource Management in Rural Australia: Issues for LWRRDC', in Mobbs and Dovers, *Social, Economic, Legal, Policy and Institutional R&D for Natural Resource Management*, p. 2.
[16] Clark, *The Policy Process*.
[17] Anderson, 'Animals, Science and Spectacle in the City', p. 29.
[18] Zuckerman, *Great Zoos of the World*, pp. 3–13.
[19] M. Brambell, 'The evolution of the modern zoo', *International Zoo News*, vol. 40, no. 7, 1980, p. 29.
[20] D. Hardy, 'Current Research Activities in Zoos', in Kleiman, *Wild Mammals in Captivity*, p. 532.
[21] Shepardson, *2nd Nature: Environmental Enrichment for Captive Animals*, p. 6; Kleiman, 'Behaviour Research in Zoos', p. 302.
[22] Shepardson, *2nd Nature*, p. 8.
[23] Benirschke, 'The Need for Multidisciplinary Research Units', p. 542.
[24] Kleiman, 'Behaviour research in Zoos', p. 302.
[25] Jeffries, *Biodiversity and Conservation*, p. 3.
[26] Wilson, 'Foreword', in Wilson, *Biodiversity*.
[27] Western, 'Conservation Biology', in Western and Pearl, *Conservation for the Twenty-first Century*, p. 31.
[28] Ibid., p. 32.
[29] C. Cussins, 'Elephants, Biodiversity and Complexity: Amboseli National Park, Kenya', Unpublished paper, Department of Science and Technology Studies, Cornell University, Ithaca, NY, 1997, p. 4.
[30] G. Caughley, 'Directions in Conservation Biology', *Journal of Animal Ecology*, vol. 63, 1994, pp. 215–44.
[31] R. Wiese, 'Is Genetic and Demographic Management Conservation?', *Zoo Biology*, vol. 13, 1994, p. 297.
[32] A. Burbidge, 'Conservation Biology in Australia: Where Should It be Heading, Will It be Applied?', in Moritz and Kikkawa, *Conservation Biology in Australia & Oceania*, p. 27.
[33] Mobbs and Dovers, *Social, Economic, Legal, Policy and Institutional R&D for Natural Resource Management*; Clark, *Averting Extinction*.
[34] Benirschke, 'The Need for Multidisciplinary Research Units', p. 540.
[35] C. Wemmer, 'Publication Trends in Zoo Biology: a Brief Analysis of the First 15 years', *Zoo Biology*, vol. 16, 1997, pp. 3–8.
[36] T. Stoinski, 'A Survey of Research in North American Zoos and Aquariums', *Zoo Biology*, vol. 17, 1998, pp. 167–88.
[37] Wemmer, 'Publication trends in Zoo Biology', p. 8.
[38] Benirschke, 'The Need for Multidisciplinary Research Units', p. 542.
[39] J. Giles, 'Conservation and Research Programs: Proposals by the Zoological Parks Board of NSW', *International Zoo Yearbook*, vol. 31, 1992, p. 1.
[40] Ibid.
[41] Kleiman, 'Behaviour Research in Zoos', p. 302.
[42] Benirschke, 'The Need for Multidisciplinary Research Units', p. 542.
[43] Kleiman, 'Behaviour Research in Zoos', p. 307.

[44] Captive Breeding Specialist Group, *The World Zoo Conservation Strategy*, p. 72.
[45] Ibid.
[46] G. Slater, pers. comm, 15 February, 2000; I. Smales, P. Menkhorst and G. Horrocks, 'The Helmeted Honeyeater Recovery Program: A View of its Organisation and Operation', in Bennett *et al.*, *People and Nature Conservation*, pp. 41–2.
[47] R. Woods, 'Malleefowl', in Australasian Species Management Program, *Regional Census and Plan*, Australasian Species Management Program, Mosman, NSW, 1998, p. 54.
[48] M. Bradley, 'Chuditch', in Australasian Species Management Program, *Regional Census and Plan*, 1998, pp. 76–7.
[49] Zoological Parks Board of NSW, Pamphlet for the Genome Resource Centre of Australia, n.d.
[50] See for example: Rabinowitz, 'Helping a Species Go Extinct', pp. 482–88; N. Leader-Williams, 'Theory and Pragmatism in the Conservation of Rhinos', in O. A. Ryder (ed.), *Rhinoceros Biology and Conservation: Proceedings of an International Conference*, 9–11 May, San Diego, CA, 1993, pp. 69–81; M. Stanley-Price, 'What Will it Take to Save the Rhino?', in Ryder, *Rhinoceros Biology and Conservation*, pp. 48–68.
[51] Bawa *et al.*, 'Cloning and Conservation of Biological Diversity', pp. 829–30.
[52] Western, 'Biology and Conservation', in Western, *Conservation for the Twenty-first Century*, p. 433.
[53] Tenner, *Why Things Bit Back*; Wenk, *Tradeoffs*; Alcorn, *Social Issues in Technology*.
[54] M. Hutchins, B. Dresser and C. Wemmer, 'Ethical Considerations in Zoo and Aquarium Research', in Norton *et al.*, *Ethics on the Ark*, p. 259.
[55] Wenk, *Tradeoffs*, p. 3.
[56] Bawa *et al.*, 'Cloning and Conservation of Biological Diversity'.
[57] See for example: E. O. Wilson, *Biophilia*, Harvard University Press, Cambridge, MA, 1984; M. E. Soule, 'The "two point five society"', *Conservation Biology*, vol. 5, 1991, p. 255.
[58] See for example: R. Noss, 'Conservation Biology, Values, and Advocacy', *Conservation Biology*, vol. 10, 1996, p. 904; Ehrenfeld, 'Preface'; H. Recher, 'Why Conservation Biology: an Australian Perspective', in Moritz and Kikkawa, *Conservation Biology in Australia & Oceania*, p. 1.
[59] McIntyre *et al.*, 'Species Triage—Seeing Beyond Wounded Rhinos', p. 604; Clark *et al.*, *Endangered Species Recovery*; R. Westrum, 'An Organisational Perspective: Designing Recovery Teams From the Inside Out', in Clark, *Endangered Species Recovery*, pp. 327–50; Mobbs and Dovers, *Social, Economic, Legal, Policy and Institutional R&D*; G. Meffe, 'Conservation Science and Public Policy: Only the Beginning', *Conservation Biology*, vol. 13, no. 3, 1999, pp. 463–4; D. Barry and M. Oelschlaeger, 'Science for Survival: Values and Conservation Biology', *Conservation Biology*, vol. 10, no. 3, 1996, pp. 905–11.
[60] R. Westrum, 'An Organisational Perspective: Designing Recovery Teams From the Inside Out', in Clark *et al.*, *Endangered Species Recovery*, p. 328.
[61] See for example: J. Ford, 'Visitor Perceptions of Zoo Animals and the Roles of Zoos', in *Proceedings of the ARAZPA/ASZK Annual Conference*, 3–7 April, Perth Zoo, Perth, WA, 1995, pp. 110–13; Anderson, 'Animals, Science, and

Spectacle in the City', pp. 27–50; Mazur, Contextualising the Role of Zoos in Conservation.

[62] See for example P. Maguire, *Doing Feminist Research: A Participatory Approach*, Centre for International Education, School of Education, University of Massachusetts, Amherst, 1987, p. 12.

4 A Vessel of Higher Learning?

[1] Editorial, The *Western Australian*, 6 August 1999, p. 12.

[2] World Zoo Organisation, *The World Zoo Conservation Strategy*, Chicago Zoological Society, Chicago, September 1993, p. 18.

[3] Kellert, 'The Educational Potential of the Zoo and its Visitor', p. 11; Mitchell, 'Conserving Biological Diversity', p. 12.

[4] See for example: Stevenson, 'Schooling and Environmental Education', pp. 69–82; Blaikie, 'Education and Environmentalism', pp. 1–20; Fien, *Environmental Education*; Pepper, *The Roots of Modern Environmentalism.*

[5] G. Hunt, 'Environmental Education and Zoos', *Journal of the International Association of Zoo Educators*, vol. 29, 1993, pp. 74–7.

[6] S. Tunicliffe, 'Zoo Education', *International Zoo News*, vol. 39/3, no. 236, 1992, pp. 15–22.

[7] Kellert, *Informal Learning at the Zoo*, p. 4.

[8] L. Williamson, 'External Funding of Special Programs', *Journal of the International Association of Zoo Educators*, no. 29, 1993, pp. 143–7.

[9] Royal Melbourne Zoological Gardens, *Education Strategy*, 1995, p. 1.

[10] Serrell, 'The Evolution of Educational Graphics in Zoos'.

[11] Kellert, *Informal Learning at the Zoo*; K. Markell, 'An interpretation planning model for use in zoos', *Journal of the International Association of Zoo Educators*, no. 29, 1993, pp. 38–45; R. L. Wolf and B. L. Tymitz, 'Studying Visitor Perceptions of Zoo Environments: A Naturalistic View', *International Zoo Yearbook*, vol. 21, 1981, pp. 49–53.

[12] T. Hohn, 'Zoo Horticulture: An Introduction and Overview', *International Zoo Yearbook*, vol. 27, 1988, p. 233.

[13] Ibid., p. 234.

[14] S. Bitgood, 'Exhibit Design and Visitor Behaviour', *Environment and Behaviour*, vol. 20, no. 4, 1988, pp. 474–91.

[15] J. Coe, 'Design and Perception: Making the Zoo Experience Real', *Zoo Biology*, 1985, p. 206.

[16] D. Hancocks, *Animals and Architecture*, Praeger Publishers, New York, 1971.

[17] Mullan, *Zoo Culture*, pp. 69–70.

[18] A. Embury, 'Royal Melbourne Zoological Gardens—Implementation and Development of the Master Plan', in *Saving Wildlife: Proceedings of the ARAZPA/ASZK Annual Conference*, 4–10 April, Currumbin Sanctuary, Queensland, 1993, pp. 91–3.

[19] Kellert, *Informal Learning at the Zoo.*

[20] Hohn, 'Zoo Horticulture', pp. 233–7.

[21] Robinson, 'Biodiversity, Bioparks, and Saving Ecosystems', p. 54.

[22] Serrell, 'The evolution of educational graphics in zoos'.

[23] Ibid.; Kellert, *Informal Learning at the Zoo.*
[24] Kellert, *Informal Learning at the Zoo*; Wolf and Tymitz 'Studying Visitor Perceptions of Zoo Environments'.
[25] S. Swensen, *Comparative Study of Zoo Visitors at Different Types of Facilities*, Yale University School of Forestry and Environmental Studies, New Haven, CT, 1980.
[26] S. Bitgood and A. Benefield, *Visitor Behaviour: A Comparison Across Zoos*, Technical Report No. 86–20, Jacksonville State University, Jacksonville, Alabama, 1987.
[27] Mazur, Contextualising the Role of Zoos in Conservation.
[28] De Young, 'Changing Behaviour and Making it Stick'; Prior, 'Responding to Threat: Psychology and Environmental Issues'.
[29] J. Hines, H. R. Hungerford and A. N. Tomera, 'Analysis and Synthesis of Research on Responsible Environmental Behaviour: A Meta-Analysis', *Journal of Environmental Education*, pp. 1–8.
[30] Prior, 'Responding to Threat', p. 14.
[31] R. Yerke and A. Burns, 'Measuring the Impact of Animal Shows on Visitor Attitudes', *Proceedings of the American Association of Zoological Parks & Aquaria*, 1991, pp. 532–9.

5 Arrangements on the Ark

[1] R. Wheater, 'Foreward: Past Progress and Future Challenges', *Edit*, vol. 5, University of Edinburgh, Edinburgh, 1994, p. xxix.
[2] Paehlke and Torgerson, *Managing Leviathan.*
[3] M. V. McGinnis, 'Myth, nature, and the bureaucratic experience', *Environmental Ethics*, vol. 16, no. 4, 1994, pp. 425–36.
[4] Paehlke and Torgerson, *Managing Leviathan*, p. 17.
[5] For example: J. F. Shogren (ed.), *Private Property & the Endangered Species Act: Saving Habitats, Protecting Homes*, University of Texas Press, Austin, 1998, pp. xi–xv, 2, 138–44; Committee on Scientific Issues in the ESA, *Science and the Endangered Species Act*, National Academy Press, Washington, DC, 1995, pp. 193–201; R. T. Simmons, 'Fixing the Endangered Species Act,' *Independent Review*, vol. 3, no. 4, 1999, pp. 511–36; J. P. Tasso, 'Habitat Conservation Plans as Recovery Vehicles: Jump-Starting the Endangered Species Act', *UCLA Journal of Environmental Law and Policy*, vol. 16, no. 2, 1997/1998, pp. 297–318.
[6] See for example: Kellert, *The Value of Life*, pp. 202–3; S. Yaffee, *Prohibitive Policy*; Clark, *Averting Exctinction*, pp. 8–9, 224–9.
[7] Kellert, 'A Sociological Perspective: Valuational, Socioeconomic, and Organisational Factors', in Clark, *Endangered Species Recovery*, p. 380.
[8] P. M. Blau, *The Dynamics of Bureaucracy: A Study of Interpersonal Relations in Two Government Agencies*, University of Chicago Press, Chicago, 1963.
[9] Ham and Hill, *The Policy Process in the Modern Capitalist State*; Paehlke and Torgerson, *Managing Leviathan.*
[10] Captive Breeding Specialist Group, *The World Zoo Conservation Strategy*, p. 30.
[11] Pfeffer, *Power in Organisations*, p. 4.

[12] Zoological Gardens Board of Western Australia, *Annual Report*, Perth Zoo, Perth, WA, 1994, p. 2.
[13] Zoological Board of Victoria, Annual Report, Melbourne, Victoria, 1991, p. 3.
[14] P. L. Berger, *Invitation to Sociology*, Penguin Books, Harmondsworth, 1979.
[15] Sackmann, *Cultural Knowledge in Organisations*.
[16] Morgan, *Images of Organisations*.
[17] See for example: J. Arnott and S. Dickenson, 'Supporting Conservation Goals through Environmental Horticulture', in *Proceedings of the ARAZPA/ASZK Annual Conference*, 3–7 April, Perth Zoo, 1996, pp. 139–43; T. Gray, 'The Cultural Diversity Performance Paradigm—Building High Performance Zoos and Aquariums', *Proceedings of the American Association of Zoological Parks and Aquaria Annual Conference*, 18–22 September, Zoo Atlanta, Atlanta, Georgia, 1994, pp. 398–403; R. La Rue, 'Working Together to Resolve the Historic Conflict Between Curators and Veterinarians', in *Proceedings of the American Association of Zoological Parks and Aquaria/Canadian Association of Zoos and Aquaria Annual Conference*, 13–17 September, Toronto, 1992, pp. 162–9; S. Pullyblank, 'Keepers and Teachers: A Potent Team', *Proceedings of the ARAZPA/ASZK Annual Conference*, 3–7 April, Perth Zoo, 1995, pp. 85–90.

6 The Corporate Ark

[1] I. Johnstone, *Fierce Creatures*, Boulevard Books, New York, 1997, pp. 37–8.
[2] P. Self, *Rolling Back the Market*, MacMillan, London, pp. 2–3.
[3] Adapted from Rees and Rodley, *Beyond the Market*; Horne, *The Trouble With Economic Rationalism*; Ralston Saul, *The Unconscious Civilisation*; Pusey, *Economic Rationalism in Canberra*; Carroll and Manne, *Shutdown*; Diesendorf and Hamilton, *Human Ecology, Human Economy*.
[4] S. Zifcak, *New Managerialism: Administrative Reform in Whitehall and Canberra*, Open University Press, Buckingham, 1994; Wettenhall, R., 'Corporations and Corporatisation: An Administrative History Perspective', *Public Law Review*, vol. 6, 1995, pp. 18–23.
[5] Rees and Rodley, *The Human Costs of Managerialism*, p. 3.
[6] J. Farrar and B. McCabe, 'Corporatisation, Corporate Governance and the Deregulation of the Public Sector Economy', *Public Law Review*, vol. 6, 1995, pp. 25–43.
[7] M. Considine, 'The Corporate Management Framework as Administrative Science: A Critique', *Australian Journal of Public Administration*, vol. XLVII, no. 1, 1988, pp. 4–18.
[8] D. Griffen, *Between Research and the Public*, paper presented at the National Museum of Natural Science, Taichung, Taiwan, 10–17 December, 1993, p. 5.
[9] Costar, *The Kennett Revolution*; Alford and O'Neill, *The Contract State*.
[10] Alford and O'Neill, *The Contract State*, pp. 1–17.
[11] J. Wanna, 'Public sector management', in A. Parkin, J. Summers and D. Woodward (eds), *Government, Politics, Power, and Policy in Australia*, Longman-Cheshire, Melbourne, 1985, pp. 61–79.
[12] Alford and O'Neill, *The Contract State*, p. 161.

[13] Costar, *The Kennett Revolution*, pp. 261–4.
[14] C. Larcombe, 'Sustainable Development of Zoological Parks and Aquaria', in ARAZPA/ASZK Annual Conference Proceedings, 17–22 April, Territory Wildlife Park, Darwin, 1995, p. 122.
[15] R. Eckersley, 'Rationalising the Environment: How Much am I Bid?', in Rees, *Beyond the Market*, pp. 237–50; Paehlke, *Managing Leviathan*; S. Rosewarne, 'Selling the Environment: A Critique of Market Ecology', in Rees, *Beyond the Market*, pp. 53–71.
[16] N. Economous, 'Corporatising Conservation: Environment Policy Under the Kennett Revolution' in Costar, *The Kennett Revolution*, pp. 193–202.
[17] Evernden, *The Natural Alien*', p. 11.
[18] R. Wilk, 'It's All Very Well, in Theory', *The Australian Higher Education Supplement*, 18 August, 1999, p. 39.
[19] Auckland Zoological Park Draft Strategic Implementation Plan 1995/1996, Auckland Zoo, New Zealand, 1995, p. 20.
[20] Ibid.
[21] Perth Zoo Business Plan 1995–2000, Perth Zoo, Western Australia, 1994, p. 22.
[22] Ibid., p. 1.
[23] Perth Zoo, Annual Report, 1996, p. 29.
[24] Ibid., p. 20.
[25] Zoological Parks Board of NSW, Annual Report 1996–1997, p. 78.
[26] Ibid., p. 102.
[27] G. Smith, I. Denney and J. Bartos, 'Toward International Best Practice in the Zoo Industry', in *Proceedings of the 48th Annual Conference of the International Union of Directors of Zoological Gardens*, Antwerp, 1993, pp. 16, 18.
[28] J. Bartos, 'Benchmarking: Best Practices in Australasian Zoos', Zoos Enriching Environments: Proceedings of the ARAZPA/ASZK Conference, 29 March–2 April, 1993, Adelaide Zoo, Adelaide, p. 53.
[29] Ibid., p. 53.
[30] Cox, *A Truly Civil Society*, p. 21.
[31] Doyle and McEachern, *Environment and Politics*, pp. 135–6.
[32] T. Meenaghan, 'Sponsorship—Legitimising the Medium', *European Journal of Marketing*, vol. 25, no. 11, 1991, p. 6.
[33] J. Zbar, 'Wildlife Takes Centre Stage as Cause-related Marketing Becomes a $250 million Show for Companies', *Advertising Age*, vol. 64, no. 27, 1993, p. 1.
[34] D. Sternbach, 'Blue Planet, Green Dollars', *Artforum*, vol. 29, no. 8, 1991, p. 17.
[35] H. Cogger, pers. comm., 1995.
[36] Zoological Parks Board of NSW, *Annual Report* 1992/1993, pp. 71, 108.
[37] J. Lukaszewski, 'The Great Crisis Management Paradox', http://www.e911.com/monos/M008.html
[38] S. Hume, 'McDonald's', *Advertising Age*, vol. 62, no. 5, 1991, p. 32.
[39] http://www.mcspotlight.org
[40] D. McEachern, *Business Mates: The Power and Politics of the Hawke Era*, Prentice Hall, New York, 1991, p. 110.
[41] Davidson, D., 'Straws in the Wind: the Nature of Corporate Commitment to Environmental Issues', in Hoffman *et al.*, *Business, Ethics and the Environment*, p. 65.

7 Sailing into Unknown Waters

[1] Dryzek, *The Politics of the Earth.*
[2] O'Riordan, *Environmentalism.*
[3] Ralston Saul, *The Unconscious Civilisation.*
[4] See for example; H. E. Daly and J. B. Cobb, *For the Common Good*, Beacon Press, Boston, 1990; Durning, *How Much is Enough?*; Diesendorf, *Human Ecology, Human Economy.*
[5] Dryzek, *The Politics of the Earth*; N. Low and B. Gleeson, *Justice, Society and Nature: An Exploration of Political Ecology*, Routledge, London, 1997.
[6] Argyris, *Knowledge for Action.*
[7] Argyris and Schon, *Organisational Learning*; T. W. Clark, 'Learning as a Strategy for Improving Endangered Species Conservation', *Endangered Species UPDATE*, vol. 13, no. 1/2, 1996, pp. 22–4.
[8] Argyris and Schon, *Organisational Learning.*
[9] Leeuw *et al.* 1994, cited in T. W. Clark, 'Appraising Threatened Species Recovery Processes: Pragmatic Recommendations for Improvements', in S. Stephens and S. Maxwell (eds), *Back from the Brink: Refining the Threatened Species Recovery Process*, Proceedings of a conference held on 12–14 December, 1995, Sydney, Australian Nature Conservation Agency, Canberra, 1996, p. 23.
[10] Morgan, *Images of Organisations*; Argyris and Schon, *Organisational Learning.*
[11] D. Ehrenfeld, 'Foreword', in Norton *et al.*, *Ethics on the Ark*, p. xix.
[12] G. Hunt, 'Foreword', in *Zoos, Evolution or Extinction: Proceedings of the ARAZPA/ASZK Annual Conference*, 15–19 April, 1996, Healesville Sanctuary, Healesville, Victoria, p. i.
[13] Morgan, *Imagin-I-zation*, p. ix.
[14] J. Elsner and R. Cardinal (eds), *The Cultures of Collecting*, Reaktion Books, London, 1994, pp. 1–6.
[15] Ibid., p. 1.
[16] Adapted from Dryzek, *The Politics of the Earth.*

Select Bibliography

Alcorn, P., *Social Issues in Technology: A Format for Investigation*, Prentice Hall, Upper Saddle River, NJ, 1997.

Alford, J. and O'Neill, D. (eds), *The Contract State*, Deakin University Press, 1994.

Anderson, K., 'Animals, Science, and Spectacle in the City', in J. Wolch and J. Emel (eds), *Animal Geographies: Place, Politics, and Identity in the Nature-Culture Borderlands*, Verso, London, 1998, pp. 27–50.

Argyris, C., *Knowledge for Action: A Guide to Overcoming Barriers to Organisational Change*, Jossey-Bass Publishers, San Francisco, 1993.

Argyris, C. and Schon, D., *Organisational Learning: a Theory of Action Perspective*, Addison-Wesley, Reading, MA, 1978.

Australasian Species Management Program, *Regional Census & Plan*, Australasian Regional Association of Zoological Parks & Aquaria, 1994–1998.

Barnes, B., *About Science*, Basil Blackwell, Oxford, 1985.

Bates, G., *Environmental Law in Australia*, Butterworths, Sydney, 1995.

Bawa, K. S., Menon, S. and Gorman, L. R., 'Cloning and Conservation of Biological Diversity: Paradox, Panacea, or Pandora's Box?', *Conservation Biology*, vol. 11, no. 4, pp. 829–30.

Beck, B., 'Reintroduction, Zoos, Conservation, and Animal Welfare', in B. G. Norton, M. Hutchins, E. F. Stevens and T. L. Maple (eds), *Ethics on the Ark*, Smithsonian Institution Press, Washington DC, 1995, pp. 155–63.

Beder, S., *The Nature of Sustainable Development*, Scribe Publications, Newham, Victoria, 1993.

Bennett, A., Backhouse, G. and Clark, T. W. (eds), *People and Nature Conservation: Perspectives on Private Land Use and Endangered Species Recovery*, Surrey Beatty & Sons, Chipping Norton, NSW, 1995.

Blaikie, N., 'Education and Environmentalism: Ecological World Views and Environmentally-responsible Behaviour', *Australian Journal of Environmental Education*, vol. 9, September, 1993, pp. 1–20.

Blau, P. M., *Bureaucracy in Modern Society*, Random House, New York 1987.

Bostock, S., *Zoos and Animal Rights*, Routledge, London, 1993.

Brown, L. R., Flavin, C. and French, H. (eds), *State of the World: A World-watch Institute Report on Progress Towards a Sustainable Society*, W. W. Norton & Co., New York, 1999.

Brunner, R. D. and Ascher, W., 'Science and Social Responsibility', *Policy Sciences*, vol. 35, 1992, pp. 295–331.

Captive Breeding Specialist Group & The World Zoo Organisation, *The World Zoo Conservation Strategy*, Chicago Zoological Society, Chicago, 1993.

Carroll, J. and Manne, R. (eds), *Shutdown: The Failure of Economic Rationalism and How to Rescue Australia*, Text Publishing Co, East Melbourne, 1992.

Carson, R., *Silent Spring*, Houghton Mifflin, Boston, 1962.

Cherfas, J., *Zoo 2000*, British Broadcasting Corporation, London, 1984.

Clark, T. W., *Averting Extinction: Reconstructing Endangered Species Recovery*, Yale University Press, New Haven, CT, 1997.

—— *The Policy Process: A Practical Guide for Natural Resource Professionals*, Yale University Press, New Haven, CT, 2000.

Clark, T. W., Reading, R. P. and Clarke, A. (eds), *Endangered Species Recovery: Finding the Lessons, Improving the Process*, Island Press, Washington D.C., 1994.

Coady, T. (ed.), *Why Universities Matter*, Allen & Unwin, Sydney, 2000.

Costar, B. and Economous, N. (eds), *The Kennett Revolution: Victorian Politics in the 1990s*, University of NSW Press, Sydney, 1999.

Cox, E., *A Truly Civil Society: 1995 Boyer Lectures*, Australian Broadcasting Corporation, Sydney, 1995.

de Courcy, C., *The Zoo Story*, Claremont, Melbourne, 1995.

De Young, R., 'Changing Behaviour and Making it Stick: the Conceptualisation and Management of Conservation Behaviour', *Environment and Behaviour*, vol. 25, no. 4, July, pp. 485–505.

Diesendorf, M. and Hamilton, C. (eds), *Human Ecology, Human Economy*, Allen & Unwin, St. Leonards, NSW, 1997.

Dixon, N., 'Protection of Endangered Species—How Will Australia Cope?', *Environmental Planning & Law Journal*, vol. 11, no. 1, 1994, pp. 6–13.

Doyle, T. and Kellow, A., *Environmental Politics and Policy-Making in Australia*, Macmillan, South Melbourne.

Doyle, T. and McEachern, D., *Environment and Politics*, Routledge, London, 1998.

Dryzek, J. S., *The Politics of the Earth: Environmental Discourses*, Oxford University Press, New York, 1997.

Durning, A., *How Much is Enough? The Consumer Society and the Future of the Earth*, W. W. Norton & Co, New York, 1992.

Durrell, G., *The Stationary Ark*, Collins, London, 1976.

Ehrenfeld, D., 'Preface', *The Social Dimension: Readings from Conservation Biology*, The Society for Conservation Biology & Blackwell Science, Cambridge, MA, 1995.

Emy, H. and Hughes, O. E., *Australian Politics: Realities in Conflict*, Macmillan, South Melbourne, 1991.

Evans, W. M. (ed.), *Interorganisational Relations*, University of Chicago Press, Chicago, 1978.

Evernden, N., *The Natural Alien: Humankind and Environment*, University of Toronto Press, Toronto, 1993.

Fiedler, P., Leidy, R. A., Laven, R. D., Gershenz, N. and Saul, L., 'The Contemporary Paradigm in Ecology and its Implications for Endangered Species Conservation', *Endangered Species Update*, vol. 10, no. 3/4, 1993, pp. 7–12.

Fien, J., *Environmental Education: A Pathway to Sustainability*, Deakin University Press, Deakin, Victoria, 1993.

Gerth, H., Wright-Mills, C., *Character and Social Structure: The Psychology of Social Institutions*, Harcourt Brace, New York, 1953.

Glacken, C., *Traces on the Rhodian Shore: Nature and Culture in Western Thought from Ancient Times to the End of the 18th Century*, University of California Press, Berkeley, California, 1967.

Graham, L. R., *Between Science and Values*, Columbia University Press, New York, 1981.

Groombridge, B., *Global Biodiversity: Status of the Earth's Living Resources*, Chapman Hall, London, 1992.

Gunderson, L. H., Holling, C. S. and Light, S. S. (eds), *Barriers and Bridges to the Renewal of Ecosystems and Institutions*, Columbia University Press, New York, 1995.

Hahn, E., *Zoos*, Secker & Warburg, London, 1968.

Hall, R. H. and Quinn, R. E. (eds), *Organisational Theory and Public Policy*, Sage Publications, Beverly Hills, 1983.

Ham, C. and Hill, M., *The Policy Process in the Modern Capitalist State*, Harvester Wheatsheaf, London, 1993.

Hamilton, G., 'Zoos and Education: It's Time (to put the words into action)', in *Saving Wildlife: Proceedings of the ARAZPA/ASZK Annual Conference*, Currumbin Sanctuary, Queensland, 4–10 April, 1993, pp. 22–4.

Hellreigel, D. and Slocum, J. W., *Organisational Behaviour: Contingency Views*, West Publishers, St. Paul, MN, 1976.

Hoffman, W. M., Frederick, R. and Petry, E. S. (eds), *Business, Ethics and the Environment*, Quorum Books, London, 1990.

Holland, K., Morton, F. L. and Galligan, B. (eds), *Federalism and the Environment: Environmental Policy Making in Australia, Canada, & the United States*, Greenwood Press, Westport, CT, 1996.

Horne, D., *The Trouble With Economic Rationalism*, Scribe Publications, Newham, Victoria, 1992.

Hunter, S. 'Overview of Biodiversity (Species and Protected Areas) Provisions', in *A New Green Agenda: The Environment Protection and Biodiversity Conservation Act*, 14 October, National Environmental Defender's Office Network, Sydney, 1999, pp. 41–4.

Jamison, A., 'Western Science in Perspective and the Search for Alternatives', in J. Salomon, F. R. Sagasti and C. Sachs-Jeantet (eds), *The Uncertain Quest: Science, Technology, and Development*, United Nations University Press, Tokyo, 1994, pp. 131–44.

Jeffries, M. J., *Biodiversity and Conservation*, Routledge, London, 1997.

Jenkins, C., *The Noah's Ark Syndrome: 100 Years of Acclimatisation and Zoo Development in Australia*, Zoological Gardens Board of Western Australia, Perth, 1977.

Kellert, S., 'The Educational Potential of the Zoo and its Visitors', *Philadelphia Zoo Review*, vol. 3, no. 1, 1987.

—— *Informal Learning at the Zoo: A Study of Attitude and Knowledge Impacts*, Report to the Philadelphia Zoological Society, 1989.

—— *The Value of Life: Biological Diversity and Human Society*, Island Press, Washington DC, 1996.

Kennedy, M., 'The EPBC Act & Biodiversity Protection: A Conservation Organisation Perspective', in *A New Green Agenda: The Environment Protection and Biodiversity Conservation Act*, 14 October, National Environmental Defender's Office Network, Sydney, 1999, pp. 45–55.

Kleiman, D., 'Behaviour Research in Zoos: Past, Present, and Future', *Zoo Biology*, vol. 11, 1992, pp. 301–12.

Kleiman, D. G. (ed.), *Wild Mammals in Captivity: Principles and Techniques*, University of Chicago Press, Chicago, 1996.

La Roe, E. T., 'Implementation of an Ecosystem Approach to Endangered Species Conservation', *Endangered Species Update*, vol. 10, no. 3/4, 1993, pp. 3–6.

Long, F., *Appropriate Technology and Social Values: A Critical Appraisal*, Ballinger Publishing Co., Cambridge, 1980.

McIntyre, S., Barrett, G. W., Kitchling R. L. and Recher, H. R., 'Species Triage—Seeing Beyond Wounded Rhinos', *Conservation Biology*, vol. 6, no. 4, pp. 604–6.

McNeely, J., Miller, K. R., Reid, W. V., Mittermeir, R. A. and Werner, T. B., *Conserving the World's Biological Diversity*, IUCN, World Resources Institute, Conservation International, World Wildlife Fund-US, The World Bank, Gland, Switzerland, 1990.

Maguire, P., *Doing Participatory Research: A Feminist Approach*, The Centre for International Education, University of Massachusetts, Amherst, 1987.

Mazur, N., Contextualising the Role of Zoos in Conservation, PhD thesis, University of Adelaide, Adelaide, 1997.

Meeks, W. W., *Beyond the Ark: Tools for an Ecosystem Approach*, Island Press, Washington DC, 1996.

Minta, S. C. and Kareiva, P. M., 'A Conservation Science Perspective: Conceptual and Experimental Improvements', in T. W. Clark, R. P. Reading and A. L. Clarke (eds), *Endangered Species Recovery: Finding the Lessons, Improving the Process*, Island Press, Washington DC, 1994.

Mitchell, G., 'Conserving Biological Diversity: A View From the Zoo', *Today's Life Science*, October, 1991.

Mobbs, C. and Dovers, S. (eds), *Social, Economic, Legal, Policy and Institutional R&D for Natural Resource Management: Issues and Directions for LWRRDC*, Occasional Paper no. 01/99, Land & Water Resources Research & Development Corporation, Canberra, ACT, 1999.

Morgan, G., *Images of Organisations*, Sage Publications, Beverly Hills, CA, 1986.

—— *Imagin-i-zation: New Mindsets for Seeing, Organizing, and Managing*, Berrett-Koehler, San Francisco, 1997.

Moritz, C. and Kikkawa, J. (eds), *Conservation Biology in Australia & Oceania*, Surrey Beatty & Sons, Chipping Norton, NSW, 1994.

Mullan, B. and Marvin, G., *Zoo Culture*, George Weidenfeld & Nicholson Ltd, London, 1987.

Nelkin, D., *Controversy, Politics of Technical Decisions*, Sage, Beverly Hills, 1979.

Norton, B. G. (ed.), *The Preservation of Species: The Value of Biodiversity*, Princeton University Press, Princeton, 1986.

Norton, B. G. (ed.), *Why Preserve Natural Variety?*, Princeton University Press, Princeton, 1987.

Norton, B., Hutchins, M., Stevens, E. F. and Maple, T. L. (eds), *Ethics on the Ark*, Smithsonian Institute Press, Washington DC, 1995.

O'Riordan, T., *Environmentalism*, Pion, London, 1981.

Ormrod, S., 'Showboat as Ark', *BBC Wildlife*, vol. 12, no. 7, 1994, pp. 40–4.

Paehlke, R. and Torgerson, D., *Managing Leviathan: Environmental Politics and the Administrative State*, Broadview Press, Peterborough, 1990.

Pepper, D., *The Roots of Modern Environmentalism*, Routledge, London, 1986.

Pfeffer, J., *Power in Organisations*, Ballinger Press, Marshfield, MA, 1981.

Primack, R., *Essentials of Conservation Biology*, Sinauer Associates, Cumberland, Massachusetts, 1993.

Prior, M., 'Responding to Threat: Psychology and Environmental Issues', *Habitat Australia*, April 1992, pp. 12–14.

Pusey, M., *Economic Rationalism in Canberra: A Nation Building State Changes Its Mind*, Cambridge University Press, Cambridge, 1991.

Rabb, G., 'The Changing Roles of Zoological Parks in Conserving Biological Diversity', *American Zoologist*, no. 34, 1994, pp. 159–64.

Rabinowitz, A., 'Helping a Species Go Extinct: The Sumatran Rhino in Borneo', *Conservation Biology*, vol. 9, no. 3, 1995, pp. 482–8.

Ralston Saul, J., *The Unconscious Civilisation*, Penguin Books, Ringwood, Victoria, 1997.

Rees, S., *Beyond the Market: Alternatives to Economic Rationalism*, Pluto Press, Leichardt, NSW, 1993.

Rees, S. and Rodley, G. (eds), *The Human Costs of Managerialism*, Pluto Press, Leichardt, NSW, 1995.

Ritvo, H., *The Animal Estate: The English and Other Creatures in the Victorian Age*, Harvard University Press, Cambridge, Massachusetts, 1987.

Robinson, J., 'Biodiversity, Bioparks, and Saving Ecosystems', *Endangered Species Update*, vol. 10, no. 3/4, p. 54.

Ryder, R., *Animal Revolution: Changing Attitudes Towards Speciesism*, Basil Blackwell, Oxford, 1989.

Sackmann, S. A., *Cultural Knowledge in Organisations: Exploring the Collective Mind*, Sage Publications, Newbury Park, CA, 1991.

Schonewald-Cox, C. M. (ed.), *Genetics and Conservation: A Reference for Managing Wild Animal and Plant Populations*, Benjamin/Cummings, Menlo Park, CA, 1983.

Self, P., *Rolling Back the Market*, Macmillan, London.

Serena, M. (ed.), *Reintroduction Biology of Australian and New Zealand Fauna*, Surrey Beatty & Sons, Chipping Norton, NSW, 1995.

Serrell, B., 'The Evolution of Educational Graphics in Zoos', *Environment and Behaviour*, vol. 20, no. 4, 1988, pp. 396–415.

Shepardson, D. J., Mellen, J. D. and Hutchins, M. (eds), *2nd Nature: Environmental Enrichment for Captive Animals*, Smithsonian Institution Press, Washington D.C., 1998.

Singer, P., *In Defense of Animals*, Blackwell, New York, 1985.

Snyder, N., Derrickson, S. R. Beissinger, S. R., Wiley, J. W., Smith, T. B., Toone, W. D. and Miller, B., 'Limitations of Captive Breeding in Endangered Species Recovery', *Conservation Biology*, vol. 1, 1996, pp. 338–48.

State of the Environment Advisory Council, *State of the Environment Australia 1996*, Department of Environment, Sports & Territories, Canberra, ACT, 1996.

Stevenson, R., 'Schooling and Environmental Education: Contradictions in Purpose and Practice', in I. Robottom (ed.), *Environmental Education: Practice and Possibility*, Deakin University Press, Burwood, Victoria, 1992, pp. 69–82.

Strauss, A. and Corbin, J., *Basics of Qualitative Research*, Sage Publications, Newbury, California, 1990.

Tenner, E., *Why Things Bit Back: Technology and the Revenge Effect*, Fourth Estate, London, 1996.

Thomas, K., *Man and the Natural World: Changing Attitudes in England, 1500–1800*, Allen Lane, London, 1983.

Vrijenhoek, R., 'Natural Processes, Individuals, and Units of Conservation', in B. G. Norton, M. Hutchins, E. F. Stevens and T. L. Maple (eds), *Ethics on the Ark*, Smithsonian Institution Press, Washington DC, 1995, pp. 74–92.

Wenk, E., *Tradeoffs: Imperatives of Choice in a High-Tech World*, Johns Hopkins University Press, Baltimore.

Western, D. and Pearl, M. (eds), *Conservation for the Twenty-first Century*, Oxford University Press, New York, 1989.

Wilson, E. O. (ed.), *Biodiversity*, National Academy Press, Washington DC, 1988.

Woodruff, D. S., 'The Problems of Conserving Genes and Species', in D. Western & M. Pearl (eds), *Conservation for the Twenty-first Century*, Oxford University Press, New York, 1989.

World Society for the Protection of Wild Animals, *The Zoo Inquiry*, 5 September, 1994.

World Zoo Organisation, *Zoo Futures 2005*, http://www.wzo.org/wzo2005.htm.

Yaffee, S., *Prohibitive Policy: Implementing the Federal Endangered Species ACT*, MIT Press, Cambridge, Massachusetts, 1982.

Zuckerman, L., *Great Zoos of the World*, Wiedenfeld & Nicholson, London, 1979.

Index

Compiled by Barry Howarth

A SCREENPLAY BY
DANIEL CLOWES
ART SCHOOL CONFIDENTIAL
PQL495319

ART SCHOOL CONFIDENTIAL, A SCREENPLAY BY DANIEL CLOWES. The script of Art School Confidential, the movie Published by Fantagraphics Books. Gary Groth and Kim Thompson, Publishers. 7563 Lake City Way, Seattle, WA 98115. www.fantagraphics.com. Printed in China. www.artschoolconfidentialthemovie.com.

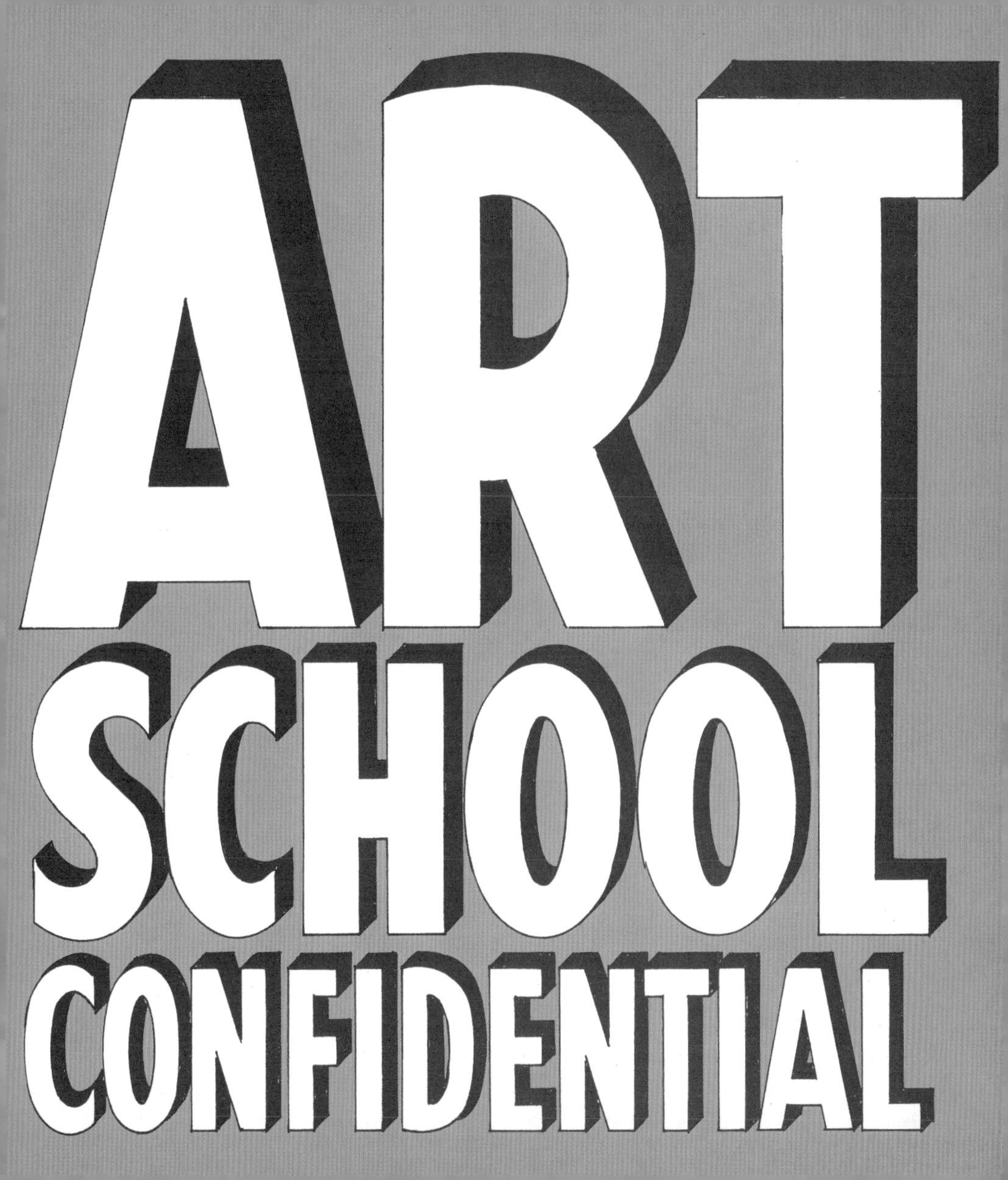
ART
SCHOOL
CONFIDENTIAL

DANIEL CLOWES: While working on the screenplay, I kept a sketchbook of character drawings to help clarify my vision of the people and situations I was writing about. I did this because I am not an act

1.

JEROME

Fairly tasteful & intentionally non-descript. Vintage dress shirts & dark pants -- a little messed-up (untucked shirt, paint spots, etc.) but not too bad.

GETS DARKER & MORE DISHEVELLED as story progresses: stubble; hair longer, more messed-up; cigs.

HAIR is greasy, unpretentious, slightly too long, but sort of "fashionable" none the less

2.

AUDREY

A little sloppy, but can't hide her innate elegance. Mostly thrift store stuff w/ one or two very expensive/fancy elements thrown in (shoes?)

Platinum Lulu Brooks hair

distinctive eye make-up & lipstick

Tries to downplay her striking beauty but dresses up more in scenes w/ Jonah (at Bushmiller's etc.)

In last scene dressed like wealthy NYC housewife.

3.

Jonah

Reads as typical college kid despite a few affectations (sideburns? Possible goatee or soul patch? earrings?)

Not so tense → oddly relaxed & comfortable in his own skin.

cargo pants

4.

PROFESSOR SANDIFORD

Could have beard

worn corduroy blazer

old faded designer jeans

Like an ex-hippie, but with a hint of femininity

NOT AS SCRUFFY as Don Baumgarten

JIMMY

Like a weatherman at first glance.

Hair weirdly perfect

er, but a mere cartoonist who could not so easily abandon the crutches of his trade.

FIRST EDITION
APRIL 2006
DESIGNED BY
THE AUTHOR
TECHNICAL
ASSISTANCE:
JOHN KURAMOTO

STILLS

BY SUZANNE HANOVER AND PETER SOREL

THIS PAGE TOP: **SOPHIA MYLES** AS AUDREY. **MATT KEESLAR** AS JONAH. **NICK SWARDSON** AS MATTHEW. BOTTOM: **JIM**

FACING PAGE: TOP: **JOEL DAVID MOORE** AS BARDO, **MAX MINGHELLA** AS JEROME. MIDDLE: **LAUREN LEE SMITH** AS BEAT GIRL. **ETHAN SUPLEE** AS VINCE. **STEVE BUSCEMI** AS BROADWAY BOB. **JEANETTE BROX** AS SHILO. **ANJELICA HUSTON** AS SOPHIE. BOTTOM: **JOHN MALKOVICH** AS PROFESSOR SANDIFORD.

THUD!